DESIGN OF
AUTOMATIC
MACHINERY

DESIGN OF AUTOMATIC MACHINERY

Kendrick W. Lentz, Jr.

VNR VAN NOSTRAND REINHOLD COMPANY

————————————————————— New York

Library of Congress Catalog Card Number: 84-3513
ISBN: 0-442-26032-6

Published by Van Nostrand Reinhold Company Limited
115 Fifth Avenue
New York, New York 10003

Van Nostrand Reinhold Company Limited
Molly Millars Lane
Wokingham, Berkshire RG11 2PY, England

Van Nostrand Reinhold
480 La Trobe Street
Melbourne, Victoria 3000, Australia

Macmillan of Canada
Division of Canada Publishing Corporation
164 Commander Boulevard
Agincourt, Ontario M1S 3C7, Canada

15 14 13 12 11 10 9 8 7 6 5 4 3 2

Library of Congress Cataloging in Publication Data

Lentz, Kendrick W.
    Design of automatic machinery.

    Includes index.
    1. Machinery, Automatic—Design and construction.
2. Automatic control. I. Title.
TJ213.L366 1985   621.8   84-3513
ISBN 0-442-26032-6

To ANNE, PHILIP, and ELIZABETH for understanding that writing a book, like designing a machine, always takes longer than you originally expect.

PREFACE

American industry's current emphasis on improving productivity has created a renewed interest in the field of manufacturing technology. Most of the recent literature, however, features the more glamorous aspects of this tech nology: Robotics, Computer Aided Design and Manufacturing (CAD/CAM), Computer Integrated Manufacturing (CIM), Expert Systems, and the "Factory of the Future." Apparently overlooked in this "new wave" are the relatively old, yet nevertheless effective, concepts of process mechanization, "hard automation," special-purpose-dedicated machines, or simply "special machines." This field (which incidentally can be just as exciting as the more esoteric pursuits) is, in fact, the one through which industry *will* realize most of its productivity gain.

When I argue that a particular manufacturing process can be "automated" much more efficiently and economically by using a special machine rather than an industrial robot, the usual rebuttal is: "Yes, but the robot can be reprogrammed at a later date to perform a different task." The important fact overlooked here is that the purchase of any piece of machinery, be it a special machine or robot, must be justified in terms of the savings that it will effect over a predetermined production run of a particular product. If the savings to be realized by hard automation of a production run justify a machine, then the fact that it may be used for some yet-to-be-determined purpose is irrelevant.

If special machinery is to be a key factor in industry's productivity gains, then we must train the designers and engineers who will design and build these machines. Technical schools and universities simply do not produce designers with this kind of expertise. One might suspect that one reason for not teaching this subject is the notion that only specific experience with a particular product can qualify one to design special machinery for processing it. There is no question that experience is a great teacher, but it is hoped that this book will serve as a first step toward generalizing the design of widely diverse machinery in such a way that technical or engineering students as well as design professionals can "learn" how to design special-purpose automatic machinery.

This book had its roots in 1979 at Novatek, Inc., a firm engaged in the design and manufacture of special-purpose automated machines and equipment. The author and Mr. Robert Metcalfe, Chief Mechanical Design Engineer, attempted to develop a "checklist" for conceiving and proposing new machines to prospective customers. The conceptual process is integral to a proposal, but prospective customers rarely pay for the creative design time spent in generating a feasible concept. Without some degree of control, the best designers find themselves spending most of their time preparing proposals and little time working on revenue-generating projects. In addition to speeding up the proposal process, the "checklist" approach also assures some degree of uniformity in the machine concepts proposed by a company.

The "checklist" approach was expanded upon by the author in an attempt to develop a training manual that would not only suggest a method for conceiving new machines but also provide specific background information and data useful to the designer in doing so. The need for such a training manual, or training program, was critical because qualified "complete-machine" designers (as opposed to machine-component designers) were (and still are) very hard to find.

Through the kind indulgence of Dr. Aldo Crugnola, Dean of the College of Engineering at the University of Lowell in Lowell, Massachusetts, and of Dr. Zelman Kamien, then Chairman of the Mechanical Engineering Department, the University has provided an excellent forum for the further development of a training program, and subsequently, this book. A total of eight courses entitled "Design of Automatic Machinery" were taught at the University by the author over the period 1981–1984. Half of the courses were taught to undergraduate Mechanical Engineering Seniors and the other half to experienced technical and engineering personnel attending the Evening Division. The presentation of material that was found to be the most effective at both levels is reflected in the organization of this book.

The book is written primarily for technical and professional people working in the fields of automation and/or manufacturing technology. Its intent is to cut across traditional disciplinary boundaries that segregate mechanical, electrical, and control engineers and to present the subject of special-purpose automatic machinery in a manner that can be understood and practiced by all.

Although problems and review questions have not been included in this edition, the book has proven useful as a textbook when supplemented by analytical examples, problems and project work. The basic organization of the book—which falls into the three major divisions of (1) actuator and work station design, (2) material transfer system design, and (3) control system design—provides an effective format for presenting the material as a college level course or in seminars oriented toward design or manufacturing professionals.

The author is extremely grateful to the University of Lowell, not only for giving him permission to develop the book through teaching of college credit courses but also for allowing him to make extensive use of University facilities including laboratories, libraries, and computer and word-processing equipment. Particular thanks are due to Prof. Frederick Bischoff, Coordinator of the Mechanical Engineering Technology Continuing Education Program, for initially accepting, then finally instituting, "Design of Automatic Machinery" as a required course in the MET curriculum.

The author would also like to acknowledge the technical input and general support offered by several persons significantly active in the field of automation technology: Mr. Dave Spaulding, President of Novatek, Inc.; Bob Metcalfe; John Six; Norm Van Dine; Jim Starbuck; and Malcolm ("Mac") Winsor.

Clearly, a book of modest length cannot begin to cover in depth all of the details of automatic machinery design as they relate to manufacturing processes in widely diverse industries. Future projects are planned that will cover the subject in greater depth, with particular emphasis on machinery and processes unique to certain industries and also on the mechanics of interfacing computers to machinery for the specific purpose of process automation.

K. W. Lentz

CONTENTS

Preface / vii

1. Introduction / 1
   1.1. Some Observations Concerning Machinery Design / 1
   1.2. The Machinery Designer / 2
   1.3. Text Objectives, Approach, and Organization / 2

2. Automatic Machinery / 4
   2.1. General Considerations / 4
   2.2. Automation / 5
   2.3. The Generalized Automatic Machine / 6
   2.4. Machine Classifications / 9
       2.4.1. Classification by Machine Function / 9
       2.4.2. Classification by Material Transfer Configuration / 16

3. The Business of Designing and Building Automatic Machinery / 19
   3.1. Design and Build of Special Machinery / 19
   3.2. The Design-and-Build Project / 21
       3.2.1. Concepting the Machine / 22
       3.2.2. The Design-and-Build Proposal / 23
       3.2.3. The Feasibility Study / 24
       3.2.4. The Preliminary Design / 24
       3.2.5. The Detail Design / 25
       3.2.6. Machine Build / 26
       3.2.7. Machine Setup, Debugging, and Proof / 26

4. The Design Process / 29
   4.1. Design and Creativity—A General Overview / 29
   4.2. System Design / 32
   4.3. A Design Method for Automatic Machinery / 34
       4.3.1. The Machine Concept Design—Obtaining Required
              Information / 36

4.3.2. The Machine Concept Design—Quantizing Process
Requirements / **38**
4.3.3. The Machine Concept Design—Identifying Alternative
Mechanizations / **43**
4.3.4. The Machine Concept Design—Synthesis of System
Concept(s) / **45**
4.3.5. The Machine Concept Design—Concept
Evaluation / **47**
4.3.6. The Machine Concept Design—Concept Selection / **49**
4.3.7. The Design Layout / **50**
4.3.8. The Design Details / **51**

5. **Machinery Economics / 53**

5.1. General Considerations / **53**
5.2. Project Costing / **56**
    5.2.1. Costing Direct Labor / **57**
    5.2.2. Overhead on Direct Labor / **59**
    5.2.3. Direct Materials Costs / **60**
    5.2.4. Other Direct Costs / **60**
    5.2.5. General and Administrative (G&A) / **61**
    5.2.6. Profit / **61**
    5.2.7. Alternative Costing Technique / **62**

6. **Actuator and Drive System Principles / 64**

6.1. Machine Forces, Torques, and Power / **64**
6.2. The Concept of a Drive System / **67**
6.3. Quantifying Machine Loads and Losses / **70**
    6.3.1. Overcoming Frictional Forces / **72**
    6.3.2. Inertial Forces in Machinery / **79**
    6.3.3. Machine Loads—Metal Forming and Metal
    Removal / **85**
    6.3.4. Forces Required to Dispense a Viscous Material / **95**
    6.3.5. Calculating Actuator Requirements from Machine
    Loads and Losses / **96**
6.4. Electric Motor Principles / **97**
    6.4.1. General Considerations / **97**
    6.4.2. AC Motors / **103**
    6.4.3. DC Motors / **110**
    6.4.4. Step Motors / **118**
    6.4.5. Servomotors / **124**

6.5. Fluid Power Actuators and Components / **128**
  6.5.1. Pneumatic Cylinders / **129**
  6.5.2. Hydraulic Cylinders / **135**
  6.5.3. Cylinder Directional Control Valves / **138**
  6.5.4. Fluid Motors / **142**
  6.5.5. Other Fluid Devices Used in Automatic
    Machinery / **144**
6.6. Solenoid Actuators / **147**
  6.6.1. Solenoid Principles / **147**
  6.6.2. Solenoid Selection / **150**

7. Practical Work Stations / 154

7.1. Selecting Off-the-Shelf Stations / **154**
7.2. Work Station Requirements / **156**
  7.2.1. Assembly Station Requirements / **157**
  7.2.2. Inspection Station Requirements / **160**
  7.2.3. Test Station Requirements / **161**
  7.2.4. Machining Station Requirements / **161**
  7.2.5. Packaging Station Requirements / **162**
7.3. Work Station Design / **163**
  7.3.1. Work Positioning / **163**
  7.3.2. Automatic Clamping / **164**
  7.3.3. Machine Accuracy / **165**
  7.3.4. Adjustment Requirements / **166**
  7.3.5. Safety Requirements / **167**
  7.3.6. Reliability and Maintainability / **169**

8. Work Transfer Subsystem Design / 173

8.1. Basic Work Flow Configurations / **173**
  8.1.1. In-line Transfer Machinery / **177**
  8.1.2. Rotary Transfer Machinery / **180**
  8.1.3. Carousel Transfer Machinery / **181**
  8.1.4. X-Y Table Transfer Machinery / **183**
  8.1.5. Work Holders / **185**
8.2. Basic Drive Principles / **186**
  8.2.1. In-line Drive System / **186**
  8.2.2. Lift-and-Carry Drive Systems / **188**
  8.2.3. Rotary Drive Systems / **192**
  8.2.4. X-Y Table Drive Systems / **196**
  8.2.5. Belt and Friction Drive Systems / **198**

8.3. Parts Handling Subsystems / 199
  8.3.1. Parts Feeders / 200
  8.3.2. Parts Orientors / 213
  8.3.3. Parts Transfer / 215
  8.3.4. Escapements / 219
  8.3.5. Parts Placement / 225

9. Machine Control System / 230

9.1. Control System Requirements / 230
  9.1.1. Types of Manufacturing Processes / 230
  9.1.2. Automatic Machinery Controls—Sequencing / 231
  9.1.3. Machine Control Requirements / 233
  9.1.4. The Control Scheme / 236
  9.1.5. Control Power Levels / 238
9.2. Relay Control Systems / 239
  9.2.1. General / 239
  9.2.2. Electromechanical Relay Construction / 240
  9.2.3. Relay Ladder Logic / 242
  9.2.4. Sequencing Controls with Relay Logic / 248
9.3. Switching Logic / 265
  9.3.1. Boolean Algebra / 265
  9.3.2. Karnaugh Mapping / 272
9.4. Electronic Logic / 282
9.5. Programmable Controllers / 285
9.6. Microprocessor and Microcomputer Control / 287

10. Sensor Principles / 290

10.1. Electromechanical Switches / 291
  10.1.1. Mechanical Contact Switching / 291
  10.1.2. Manual Switches / 296
  10.1.3. Limit Switches / 296
  10.1.4. Reed Switches / 301
  10.1.5. Pressure Switches / 302
  10.1.6. Electromechanical Timers / 303
10.2. Solid State Switches / 305
10.3. Proximity Switches / 306
10.4. Photoelectric Switches and Controls / 308

10.5. Rotary Position Sensors / **311**
    10.5.1. Rotary Encoders / **311**
    10.5.2. Analog Angular Position Sensors / **315**

Appendix 1 / **319**

Index / **331**

1. INTRODUCTION

1.1. SOME OBSERVATIONS CONCERNING MACHINERY DESIGN

In 15 years as an engineer, manager, and entrepreneur in the fields of mechanical, electromechanical and thermal systems design I have made the following observations:

- Relatively few engineers and/or designers get the opportunity to design, see built, and render operational a *complete functional machine.*
- Courses and texts which treat the subject of designing and building automatic machinery in a comprehensive manner are relatively scarce.
- As a result, very few engineers and/or designers have backgrounds—either by experience or by education—in the specific combination of disciplines necessary to design and build a complete automatic machine.
- Competent engineers will readily undertake design assignments for components, some extremely complex, yet they hesitate to accept responsibility for designing a complete machine, no matter how simple and straightforward that system might be.
- The most effective machine design is done by a single individual who takes the machine from concept to detail design. The design is internally consistent and represents the "best" selection and integration of many alternative approaches.
- The top notch machinery designers cannot concisely explain "how they do it."

The conclusion one would draw from these observations is that design of automatic machinery is still more of an art than a science, even in this day of microprocessors, computer controls, CAD/CAM, and industrial robots. This was, in fact, the motivation for writing a book which attempts to understand and quantify the art, and to present a formal design process along with appropriate engineering fundamentals.

1

1.2. THE MACHINERY DESIGNER

Traditionally, the complete-machine designer evolved from a board designer who, by fortuitous combination of skill, self-drive, and opportunity, took on more and more responsibility for segments of the overall machine design, until he acquired total machine design responsibility. Such designers were, and still are, hard to come by, extremely valuable to their employer, and extremely difficult to replace when they leave. They develop the art in a practical, hands-on manner which is difficult to quantify and teach. It should come as no surprise, then, that the seeds for this book were planted in 1980 when, as Vice President of Operations for a special machinery design-and-build company, I attempted to devise a method for quickly and efficiently training promising young design engineers to be top notch complete-machine designers.

Perhaps the greatest weakness of traditional machinery designers is in the area of automatic controls. Because machine control systems, particularly electronic or computer controls, are so adaptable and able to meet most any control requirement, the machine is usually designed first and the controls added or applied later. It is fairly common industry practice that the machine be designed first, then "the electrical guys called in" to handle the controls.

This set of circumstances has led to specialization or polarization in machinery design which results in a sharing of the total design responsibility between mechanical and electrical specialists. Consequently one finds (although not in great numbers) excellent mechanical design engineers who create machines of great precision, innovative function, and superior reliability paired with the electronic engineer proficient in circuit design and/or computer programming who is not expert in, nor often interested in, the mechanical details.

Considering the scarcity of (good) individuals of either type, it is of little surprise that the top-flight machinery designer embodying the talents of both types is virtually nonexistent. It is indeed rare to find an individual who can conceptually design an entire machine, from nuts and bolts to microprocessor controls, and moreover do it quickly, with a high probability that the machine, once built, will work as designed. Hopefully this book will contribute in some small way to the integration of machinery design and machinery controls technology in a manner useful to the designer of complete machinery systems.

1.3. TEXT OBJECTIVES, APPROACH, AND ORGANIZATION

Even the most experienced machinery designer will admit that it is not so much his vast recollection of facts and figures, but his overall approach to machinery design problems which contributes most strongly to his success. With this in mind, we attempt to postulate a *design process* which is hopefully general enough to apply to all of the many types of automatic machinery possible, yet specific enough to guide the novice step by step through all aspects of the machine design. Chapter 4 discusses design methods in general, and then develops a design process tailored specifically to the design of practical automatic machines.

It has already been pointed out that even the most competent engineers balk at undertaking complete-machine design tasks as opposed to machine-component design tasks. This is primarily a matter of confidence (or lack thereof) in forging ahead from concept to completed design. It is therefore one of the objectives of this text to promote individual confidence in designing complete systems, not only by providing a specific method, but also by emphasizing that there are many possible solutions to the same design problem. As long as sound principles are used in designing that machine, no one can criticize that design simply because it is "different."

As with all endeavors, experience is a very important factor. A third objective of this book is to develop awareness or "pseudo-experience" regarding specific off-the-shelf components and subsystems available for incorporation into machines. Certainly individual confidence is bolstered if one is familiar with what is available in the way of hardware. Chapter 3 provides background into the nature of the design and build business and typical design and build projects. The technical chapters, 5 through 10, refer extensively to specific as well as generic types of machinery components available to the designer. Where appropriate, actual case studies of machines with which the author has been involved are cited.

A fourth objective is to provide general understanding of a thinly documented and rarely taught field of mechanical engineering. An attempt is also made to portray the design-and-build (of special automatic machinery) business, which itself is not well known.

The final and perhaps the most academic objective is to synthesize a great many diverse engineering fundamentals into a practical and applicable field of its own—design of automatic machinery. Chapters 5 through 10 summarize the various engineering principles encountered in machinery design.

2. AUTOMATIC MACHINERY

2.1. GENERAL CONSIDERATIONS

The subject of this text is automatic machinery, but it should further be pointed out that it deals primarily with *special production machinery*, defined in a broad sense.

Special implies that the machine is one-of-a-kind and has a specialized function in terms of the operation or series of operations it performs on a particular product. *Production* implies that the machine performs a repetitive function, presumably (but not always) on large numbers of product items at as-fast-as-possible rates. The two terms are often used in concert, since it is difficult to justify on an economic basis designing and building a special purpose automatic machine (very expensive, almost by definition) unless it results in substantial production cost savings. Production cost savings are not usually recognized unless the product is made in large quantities at high rates. However, special automatic machines may be (and often are) built which perform their function at relatively slow rates. Such performance is justifiable when the machine provides the *only* way of producing a product or of performing a particular function. The same holds true of a machine which affords a level of quality or reliability unobtainable by manual means.

Industrial robots are not themselves special machines because they are mass produced. Their programming "specializes" them for particular applications. Indeed, a special automatic machine might include one or more industrial robots as system components. Although robot manufacturers would probably disagree, it is doubtful that industrial robots will ever replace special machines as the most efficient manufacturing medium. By the time a robot has been custom programmed, fixtured to interface with a particular product and provided with required auxiliary functions such as input product feed, in-process inspection, product packaging, etc., it has become a "special machine," and quite possibly the same end result could have been accomplished less expensively without the robot.

As a matter of clarification, the term *production* is meant to include any machine function which is performed repetitively on a product or object which is not part of the machine. Thus, this text deals with the design and building of automated special purpose machinery systems capable of such diverse functions as assembly, machining, inspection, testing, packaging, or simple material transport. Operational rates may be high or low. The common function is a repetitive sequence of operations automatically performed on a series of objects or products.

2.2. AUTOMATION

A great deal has been written in an attempt to precisely define *automation* and to categorize machinery which is more or less automatic. In all cases, the degree of automation ends up being entirely arbitrary. Fortunately, semantics are not of great importance in engineering design. What matters is not whether the machine is automatic or semi-automatic, but whether or not the machine works the way we want it to work. The purely arbitrary definitions used in this text, which seem to be fairly well accepted industry practice, are as follows:

Manual or Manually Operated Machine. A machine in which both control and power are initiated by the machine operator, although power may be multiplied or precisely directed by the machine. Examples include hand tools; hand power tools; arbor press; single station pneumatic press (actuated by operator); self-powered, hand held screwdrivers; and standard drill press.

Semi-Automatic Machine. A machine which performs one or more sequential operations without operator input, i.e., automatically, but which requires operator control inputs somewhere in the work performance sequence other than to start or stop the machine. Examples include a garage lift, tool room lathe (with power feed), milling machine (with power feed), manually fed assembly station.

Automatic Machine. A machine which performs a repeating sequence of operations on a continuous flow of workpieces without operator input(s) (except START, STOP, EMERGENCY STOP, etc.). The machine automatically controls work flow and positioning and can theoretically run unattended until out of input workpieces. Examples include automatic feed screw machines, magazine or bowl fed assembly machines, and computer controlled DIP insertion machines. The machines are defined as automatic, even if the magazine or vibratory bowl feeder is hand filled. It is the actual performance of operations which we consider as the determinant of machine classification, not the nature of its setup.

Adaptive Machine. An automatic machine is said to be adaptive if it includes control features which permit it to make internal adjustments in response to changing environment or machine characteristics so as to assure continuing proper output of product. In its most basic form, an adaptive machine must sense product variation, decide what must be done to correct the problem, then automatically implement that correction. Examples of adaptive features include automatic tool sharpening and advancing, automatic adjustment of bottle fill quantity in response to liquid level inspection, and calculation of material removal requirements for input material of varying composition and properties.

It should be pointed out that *automatic* does not necessarily mean computer, microprocessor, or even electronic control. It simply means that a physical operation or series of operations is performed by the machine without human input.

2.3. THE GENERALIZED AUTOMATIC MACHINE

This text adopts a generalized automatic machine concept which arbitrarily divides all machines into three subsystems: work stations, material transfer, and control. This generalization originates from the traditional concept of an automated production line or multistation production machine, wherein a continuous flow of work is positioned at sequential work stations and some operation performed on the workpiece at each location. At any given time, therefore, there are as many workpieces in process as there are work stations, each at a progressively higher level of completion.

All modern machines do not, obviously, fit the traditional model; however, all modern automatic machines *can* be broken down into the three subsystems for purposes of design. The design process developed and discussed in detail in Chapter 4 uses this arbitrary breakdown as a guide to defining machine requirements. Machinery component selection is also facilitated by the three-subsystem breakdown, since work stations, material transfer devices, and control systems are available in various configurations as "off-the-shelf" packaged subsystems.

Fig. 2-1 depicts the generalized machine graphically. There are n work stations, connected by the mainline material transfer mechanism(s). To be completely general, crossline material transfer, or parts feed, is associated with each station, although this function is usually associated with assembly machines as opposed to machine tools, inspection machines, or test machines. In packaging machinery, the packaging material itself can conveniently be thought of as "parts", since material other than the work must be fed to the station, then attached to the product.

Work Stations. The work station is the machine subsystem which actually performs one or more operations on the workpiece. In a machine where the

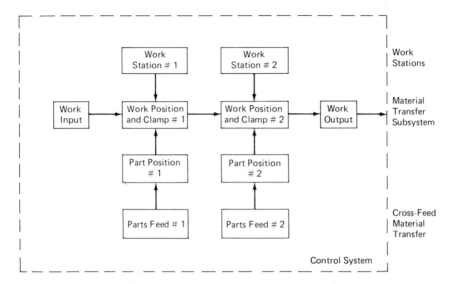

Fig. 2-1. The generalized automatic machine.

work actually stops at a location (i.e., an intermittent machine), the work station is easily visualized as the machinery at one point in the machine. If, however, the operation is performed on the workpiece as it is moving (i.e., in a continuous machine) then the work station is difficult to associate with a location, although the identification of the work station machinery should not be difficult.

In general it is good design practice to use a "packaged" work station, that is, an off-the-shelf, standard product capable of performing the desired operation. The unit need only be bolted to a table and wired into the control system in order to become a work station. Examples include milling or drilling heads, riveting heads, adhesive application heads, heat sealing heads, spray heads, laser scribing heads, ultrasonic welding heads, soldering heads, etc. Table 2-1 lists the advantages and disadvantages of selecting packaged work stations as opposed to designing the desired work station "from scratch".

There are many machinery requirements for which standard off-the-shelf packaged work stations are not available. In this case there is little choice but to design it oneself. Here we must select the proper actuator as the basis for the work station, and then tool it appropriately. The most common work station actuators are electric motors, pneumatic and hydraulic cylinders, and solenoids. The basics of these devices are discussed in detail in Chapter 6.

Because finished product quality depends very strongly on the relative positioning of the work with respect to the work head, the function of work

TABLE 2-1.   Selecting Off-the-Shelf Machinery components vs. Designing From Scratch.

Advantages to Selection of Existing Packages	Disadvantages to Selection of Existing Packages
Packaged components have been fully engineered and de-bugged.	Possibility exists of compromising machine performance down to the level of the existing package.
Packaged components have been field tested and have probably undergone design revisions.	Existing packages are not always aesthetically complementary to your machine.
Component reliability has been demonstrated through actual operating experience.	It may be necessary to incorporate additional stations or operations in order to use an existing package.
Manufacturer warantees the component or system.	
Large scale production results in lower cost vs. fabrication of one item of your own design.	
Manufacturer sales or engineering personnel can be extremely helpful in solving your specific problem.	

position and clamping is here defined as part of the work station. This definition is arbitrary, since the work positioning and clamping function could also logically be associated with the material transfer subsystem. In practice, however, there will almost always be a control algorithm in which work positioning and clamping will be intimately related to work station operation. That is, the work station will not actuate until the presence of correctly located and clamped work is verified and a control signal generated.

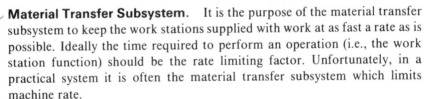

Material Transfer Subsystem.   It is the purpose of the material transfer subsystem to keep the work stations supplied with work at as fast a rate as is possible. Ideally the time required to perform an operation (i.e., the work station function) should be the rate limiting factor. Unfortunately, in a practical system it is often the material transfer subsystem which limits machine rate.

Work can be moved through the machine with or without work holders, or pallets, and the path taken from input to output by the work is referred to as the *mainline* material transfer route. It is usually either *inline* or *rotary* in configuration, and indeed, the main line material transfer configuration serves as one method of categorizing machines.

The material transfer subsystem also includes the mechanisms for feeding, orienting, transferring, and placing parts onto the workpiece as it moves through the machine. These functions are referred to here as the *crossline* material transfer route. *Parts* is a general term, and can refer to any material applied to the workpiece, including adhesives, paint, packaging film, or even product itself, as in the case of feeding liquids into bottles.

Control System. The control system is the machine subsystem which coordinates work station actuation with work transfer between stations. The primary function of the control system of an automatic machine is to synchronize mainline work flow (transfer, positioning, and clamping) with crossline work flow (parts transfer and placement) and with work station operation.

It is difficult to place the control system on a block diagram. There will always be a control output to each work station telling that device when to actuate. Depending on the degree of sophistication of the machine controls, there may be sensors to determine work or parts presence, or sensors to determine work speed or position during transfer. The actuator(s) which power the material transfer system(s) are most generally the devices under control and synchronization by the overall system. Fig. 2-1 depicts the control system of the generalized machine as an all-encompassing (dotted line) enclosure to demonstrate the impact of the control system on all machine subsystems.

2.4. MACHINE CLASSIFICATIONS

One often hears a machine referred to as an "inline assembly machine," "rotary assembly machine," "computer controlled machine," etc. As is the case with most large groups of generically similar equipment, the classification of a particular device can be made in any one of a number of ways, and is usually very arbitrary. Automatic machinery is no different in this respect, and the preceding discussion of a generalized machine forms a very convenient basis for arbitrarily classifying machine types. Generally speaking, machines are classified by function of the principal work station, by method of mainline work transfer, or by the predominant method of control.

2.4.1. Classification by Machine Function

Machinery may be arbitrarily classified into five basic types which describe the operation performed by the machine. These general types are:

- Assembly machines,
- Inspection machines,

3 ● Test machines (or equipment),
4 ● Packaging machines,
5 ● Machine tools.

The machine may be single station or multistation, in which case there may be a mix of work station types. In general, the multistation machine is referred to in terms of the function of the predominant work station.

1 **Assembly Machines.** _Definition._ A machine which assembles and fastens together two or more separate items of material (parts) in a predetermined sequence.

In the special machinery design-and-build business, assembly machines are probably the most common. This is because assembly of most products is basically different from assembly of all other products by virtue of the product itself. Thus there is a greater need for designing custom assembly machines than there is for custom packaging machinery or machine tools, since these types of machines are more easily standardized.*

Fig. 2-2 depicts the generalized operational breakdown of an assembly station. The first requirement is to position the workpiece with respect to some machine reference. This positioning may be inherent in the material transfer subsystem, or it may require a specific action, such as clamping the workpiece against a V-block reference. The object of the positioning step is to provide a precise target for the assembled part which must be located with respect to the workpiece. Step 2 then is to feed the assembled material, or part, into, onto, next to, etc. the workpiece for subsequent (or simultaneous) fastening. The part must have the proper orientation as well as location with respect to the work.

The third step is one of fastening the crossline-fed part to the mainline workpiece by any one of many possible methods. Note that steps 2 and 3 may very well be a single operation, in that placement of the part with respect to

---

*It is interesting to note that products are _always_ designed under the constraints of existing fabrication or machining methods. As a result, one will never find a part which cannot be made by one of several standard fabrication techniques.

In a similar vein, one rarely finds a need for an exotic packaging technique, probably because packaging of a product is usually not critical to the manufacturing process. There are usually several packaging options, and a fairly standard packaging machine (station) will probably fulfill at least one option.

Rarely is a product originally designed (they are often redesigned) under the constraints of existing automatic assembly techniques or machinery. Assembly machinery is usually custom designed to assemble existing parts. The predominant reason is that many products are not originally meant to be assembled automatically, so automation is an after-the-fact occurrence. A more subtle reason is that a well defined set of automatic assembly constraints does not exist for product designers to use.

1. Position work A relative to machine reference.
2. Feed material (part) B, to A in proper orientation.
3. Fasten B to A.
4. Inspect A/B assembly. (Optional)
5. Repeat cycle.

Fig. 2-2. Operational requirements of an assembly station.

the work also constitutes fastening. This would be the case for inserting a DIP into a socket, inserting a part designed for a snap fit, or for placing a part onto a previously applied adhesive surface. (Note that in the case of adhesives, a previous assembly operation must be assumed in which the adhesive was a part.)

These three steps constitute the assembly operation in its most fundamental sense, but there are also some often used optional steps which greatly enhance machine reliability. First, it is good machine design practice to provide a means of sensing workpiece presence, preferably presence with proper reference location. Until this is done, the machine should prevent step 2 (part placement).

Just as it is wise to check workpiece presence, so it is also good practice to sense part presence, and if important, part orientation. This precaution will prevent the machine from continuing to "fasten" no parts to the workpiece. Once workpiece presence and part presence are sensed, the actual fastening step may then take place.

The resulting assembly may be the final output from the machine, or it may be a subassembly which is moved by the material transfer system to a following station for subsequent assembly operations. In either case it is wise to inspect the assembly to assure that it in fact took place as desired. Inspection may be carried out at a separate inspection station, or the inspection device may be incorporated directly into the assembly station. Use of interim assembly inspection steps (or separate stations) represents decisions which are usually made on economic considerations. The basic question is whether the cost savings to be realized by preventing faulty assemblies justifies the extra cost of inspection devices, inspection stations, and control systems of increased intelligence. The cost savings in turn depend on:

- The probability of misplacing the part with respect to the workpiece,
- The probability of misfastening the part to the workpiece,
- The cumulative value of the assembly at the level of inspection.

Inspection Machines. *Definition.* A machine (or station) which inspects a component, subassembly or assembly for a measurable physical property (e.g., dimension, weight, magnetic properties, electrical properties, etc.).

1. Position target feature of work A relative to sensor.
2. Perform measurement by:
   a. Traversing A past sensor and actuating measurement system, or
   b. Traversing sensor past A and actuating measurement system, or
   c. Without relative motion, actuating measurement system.
3. Transmit measurement data:
   a. To machine, for control, and/or   with or without intermediate
   b. To operator display, and/or   signal conditioning.
   c. To storage device
4. Repeat cycle.

passive

Fig. 2-3. Operational requirements of an inspection station.

An inspection machine or an inspection station within a larger machine requires, by definition, a sensing device to measure the physical property in question. Fig. 2-3 shows the generalized operational breakdown of an inspection station. Perhaps even more important than in an assembly machine, the first step of positioning the workpiece (object of inspection) with respect to the machine (or the sensor) is critical. Positioning is essential to accurate measurement with most sensing devices, particularly noncontacting types.

The second step is that of making the actual measurement. The possible methods are:

- Traverse work past sensor and actuate measurement system.
- Traverse sensor past work and actuate measurement system.
- Provide no relative motion and actuate measurement system.

The third step is that of transmitting measurement data so derived:

- To the machine as internal feedback for control purposes,
- To the operator, by means of a display device,
- To a data storage device.

In practice, inspection data may well be used in all three modes simultaneously.

As was the case with assembly machines, the option exists to sense presence of an object (workpiece) and then use this information to allow the actual operation, in this case, measurement, to take place. This will avoid taking and storing meaningless data because a workpiece was not present, or was improperly positioned.

3 **Test Machine (or Equipment).** *Definition.*—A machine which tests a subassembly or assembly for *performance* against a specification. Perfor-

mance is generally the response of the assembly to a calibrated input which itself is part of the machine.

A test machine is not an inspection machine because it is active, in the sense that it stimulates the object (workpiece) by providing an input of some type, and senses the object's response to that input. An inspection machine, on the other hand, is passive, in that it only measures a property inherent in the workpiece. There is a very large field known as Automatic Test Equipment (ATE) which pertains primarily to electrical or electronic components. The input to the workpiece, in this case electronic devices, is electrical, as is the measured output. The type of test machinery we are concerned with here consists of those machines which provide a mechanical input to an object (workpiece) and measure the response to that input. The measured output is often electrical in nature, as with automated equipment to test transducer output for given mechanical inputs (e.g., pressure, temperature, acoustic excitation, rotation, etc.)

The test machine must first position the object with respect to both the input device(s) and the output sensors. The problem is identical to that of an assembly machine, however, since both input and output measuring devices are referenced to a fixed point on the machine.

The second step, not found on an assembly machine, is to access both input and output ports of the object by the machine. To provide a pressure input to a pressure gage, for example, a pressure supply tube might advance and form a seal against the object gage, mating with its pressure input port. To read an electrical output, a set of contacts might advance from the machine and make contact with the corresponding output leads of the object.

The third step is to actuate the input stimulus, and the fourth step is to read the appropriate response. As in the case of an inspection machine, the data, once read, can be:

- Transmitted to the machine as internal feedback,
- Transmitted to the operator by means of a display,
- Stored on tape or in computer memory.

Fig. 2-4 depicts the operational breakdown of steps which take place in a single cycle of an automatic test machine or test station.

Packaging Machine.   _Definition._ A machine which results in the workpiece being completely encapsulated within a package or packaging material. The object is bagged, wrapped, tied, boxed, bottled, etc.

Packaging machinery is a specialized and relatively standardized field within the design-and-build industry. Design of special packaging machinery does not usually involve innovative functional design as much as it involves

1. Position object A relative to input device(s) and output (performance) measuring sensors.
2. Access input, output ports.
3. Actuate input stimulus.
4. Measure performance response. *Active*
5. Transmit performance data:
   a. To machine, for control, and/or
   b. To operator display, and/or
   c. To storage device.
6. Repeat cycle.

Fig. 2-4. Operational requirements of a test station.

adapting standard techniques or machines to unique objects. For cases where an off-the-shelf packaging station cannot be used in a machine, the following steps define the generalized station operation (See also Fig. 2-5).

The first step, as in an assembly station, is to position the object with respect to the package or packaging material. In a manner of speaking, packaging is assembly of an object into a package.

The second step is to actuate the packaging action, motion, or motions required. This may include insertion, deposition, or dispensing of a premeasured amount of object material into a pre-prepared package, or physically wrapping the packaging material about the object, or vice versa (i.e., move the object so as to wrap it within the packaging material). Note that this step of actuating the packaging action can be very complex and may require more than one single operation. In particular, an object may have to be rotated about more than one axis to complete the wrapping motion.

The third step is to actuate package closure. This can be done by feeding and applying a separate part, such as capping a bottle, or it can be done by folding and taping package flaps, as in closing a cardboard carton. Objects encapsulated by plastic film or inserted into plastic bags may be closed by heat sealing.

1. Position object A relative to package or packaging material B.
2. Actuate packaging action or motion:
   a. Insert, deposit, or dispense (premeasured) A into B; or
   b. Wrap B about A.
3. Actuate package closure:
   a. Feed and close with separate part (capping).
   b. Seal package. (Sever packaging material if continuous.)
4. Perform auxiliary functions (optional):
   a. Labeling.
   b. Counting,
   c. Sorting.
5. Repeat cycle.

Fig. 2-5. Operational requirements of a packaging station.

Optional packaging machine or packaging station steps include labeling, counting and sorting. Off-the-shelf components are available for labeling and counting. Sorting usually requires custom programming of the machine control system to sort according to a very specific, user-defined algorithm.

Machine Tool.   _Definition_. A machine which physically alters the workpiece by removing or by rearranging existing material.

Machine tools are even more standardized than packaging machinery. The primary reason is that there are several well defined machine tool types which are capable of producing a very large number of possible shapes. Those shapes not attainable with standard machine tools are generally not designed into products.

Nevertheless there is a need to design special machinery capable of performing machine tool functions such as metal removal or forming. Often the design requirement is for single station of a much larger machine where a specialized machining operation is required. Fortunately, the common machine tool operations can be obtained as off-the-shelf modules which are easily integrated into the special machine. Examples include milling heads, drilling heads, pressing or stamping heads, and automatic feeds for work or for tools.

The need often arises for work stations which cut, stamp out, or bend part of a workpiece or fed part as part of an assembly operation. It is often more efficient to do light machine tool operations on line rather than try to feed premachined parts. The concept of making a part to order as it is required by an assembly is one which results in fewer misfeeds, fewer misfits, therefore fewer rejected parts or less machine downtime.

Fig. 2-6 shows the operational steps involved in a typical machine tool work station. As in all other machine types, the first step is to position the work with respect to a machine reference, usually the tool which will be doing the machining operation. The second step is to then perform the machining operation, by moving the work with respect to the tool, or by moving the tool with respect to the work. Tool or work feeding may be a required part of the machining operation.

1. Position work A relative to tool B.
2. Perform machining operation.
   a. Move A relative to B, or
   b. Move B relative to A.
3. Remove waste material (chips).
4. Maintain nominal conditions (i.e., coolant).
5. Inspect fabricated feature(s) (optional).
6. Repeat cycle.

Fig. 2-6. Operational requirements of a machine tool station.

In the case of material removal, provision must be made for disposal of waste material (chips). A workpiece cooling system must also be provided if the operation results in significant heat generation. As is the optional practice in assembly machinery, an inspection step may also be provided in a special machining operation.

2.4.2. Classification by Material Transfer Configuration

When multiple operations must be performed on a workpiece, the workpiece must move through the machine so it can be operated upon by each of several actuators or tools at the work stations. In general, there will be zero relative motion between the workpiece and the work station actuator for a finite period of time during which the operation is performed. Zero relative motion between work and work station is achieved by stopping the work at a point (*intermittent transfer*) or by matching the work speed to that of the workhead (*continuous transfer*). Of course, there are examples of work moving relative to the workhead at a given work station during an operation. Examples include "on-the-fly" cutoff, milling, non-contact dimensional inspection, and seam welding.

The general layout of work stations with respect to the path that the work follows as it moves through the machine forms the basis for another arbitrary method of classifying machine types. Machines in which the work moves generally in a straight line are referred to as *in-line* machines. (see Fig. 8-1). Machines in which the work moves in a circle, coming into place beneath workheads located about the periphery of a round table, are known as *rotary* or *dial* machines (see Fig. 8-4). Machines in which the work follows a combined linear and rotary path as described by a chain about two sprockets are called *carousel*-type machines, and in fact are often chain driven (see Fig. 8-5). Fig. 2-7 summarizes the various families of machinery as classified by material transfer configuration.

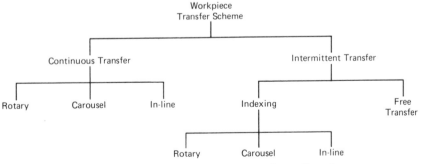

Fig. 2-7. Machinery classified by work transfer configuration.

Continuous Transfer Machinery.   In a continuous transfer machine, the work moves at a constant velocity, as do the workheads, so as to maintain zero relative velocity between workpiece and workhead. For obvious reasons, these machines are also called *synchronous* machines.

Continuous transfer machinery offers the highest unit cycling rate (greater than 6000 pieces per hour) because the work does not have to start then stop for an operation to be performed. To maximize production rate, however, multiple heads must be used. In general, the workheads will be relatively expensive, so multiple workheads make the continuous machine rather costly.

Although it is conceivable that an in-line machine might be continuous, with workhead accelerating to match work speed, then quickly returning to pick up the next unit, this is impractical because of the great amount of power required for workhead acceleration and deceleration. Common continuous machines are rotary, with workheads rotating in synchronism with rotary fed work. A good example is a bottling machine, where filling and capping is done as bottles (and filling heads) move continuously.

Advantages to using a continuous machine include the following:

- High production rates are possible.
- Lighter overall machine construction is possible because intermittent starting and stopping is not required, so vibrations are not a serious problem.
- Continuous controls are easier to design because sequencing is not required.
- Workheads and transfer mechanism may be driven by the same motor.
- Continuous motion allows integration of actions (i.e., in rotating).

The following rules apply to the selection of a continuous machine concept:

- Tooling or workhead should be simple and lightweight (low inertia) relative to workpiece.
- Tooling or workheads should be relatively inexpensive if several are to be used.
- Continuous machinery should be used when the work is not amenable to intermittent starts and stops (e.g., liquid or soft food packaging, low inertia work with "fly-away" tendencies).
- It should be used when high production rates cannot be achieved otherwise.

Intermittent Transfer Machinery.   By far the most common type of automatic machinery is of the intermittent rotary or in-line type. (Excluding continuous process equipment such as fabric or film drawing, drying, etc., which is not the subject of this book). Here the workpiece moves from one station to the next, with one or more operations performed at each stop. In an *indexing* machine, the machine cycling rate is limited to the slowest operation performed, since the work indexes at a constant rate. Generally all workpieces are driven from the same indexing actuator.

In a *free transfer* machine, the work moves intermittently, but not necessarily at a single constant rate for all stations. There are buffer zones between stations so that work accumulates ahead of a slow cycling station, and depletes the buffer stock ahead of a faster cycling station. Free transfer machinery is also referred to as *flexible* machinery, and the concept applied in general to a large manufacturing system is called *flexible manufacturing*.

When an indexing machine experiences a jam-up at one station, the entire machine must be stopped while that station is cleared. If one station of a flexible system goes down, the system can continue to function using buffer stocks while the jam is cleared.

In any type of machine, the workpiece may be transferred through the machine in two ways:

1. Directly, by sliding the work along, or by lifting and carrying the work down the line. Here the workpiece must be designed for direct transfer by having a positioning or orienting feature, and by having suitable surfaces for sliding or lifting.
2. By means of a pallet or work carrier (also called *nests* or *fixtures*) which holds the work throughout the machine. The work holders must recycle after carrying a workpiece completely through the machine.

Chapter 8 discusses the design of material transfer subsystems in some detail.

3. THE BUSINESS OF DESIGNING AND BUILDING AUTOMATIC MACHINERY

3.1. DESIGN AND BUILD OF SPECIAL MACHINERY

The design-and-build business is one populated primarily by relatively small firms, most of which were first machine shops which then expanded into design houses. By necessity they are concentrated in industrial manufacturing areas, where they offer local manufacturers a wide range of services including: machining components to customer's prints; fabricating and assembling tools, fixtures, or complete machines to customer's prints; or designing, building, and installing a complete machine to meet customer production requirements.

It is my observation that the evolution of a basic machine shop into a design-and-build house is one of progressively taking more and more of the design responsibility from the manufacturing specialist by a relatively small, efficient, independent machine builder who can react more quickly than the specialist and pay more attention to minute detail. The independent machine builder has the flexibility to make design changes as the machine is being built, a process which is very cumbersome to a production facility which does not have an in-house equivalent of a design-and-build shop. Clearly, the business of a manufacturer is to produce his product, not to design production machinery. The existence of machine design-and-build specialists who can come into his plant, assess the requirements for a special purpose production machine, design and build that machine, and install a properly functioning machine with minimal impact on existing production, makes life much easier for the manufacturer.

To be completely fair, it must be pointed out that not all design-and-build houses are small, and not all special machine design and build is done by independent specialists. However, these exceptional outfits face some fundamental problems which are by no means few in number.

The large machine design-and-build house faces the problem of maintaining a large volume of business in order to keep its staff busy. As the business grows, it becomes more difficult to handle a great number of different machines, and inevitably, the firm begins to look for a standard product. As standard products are obtained and successfully integrated into the business, the specialty work becomes less attractive, at least from a pure business point of view. The problems arise, in my opinion, because the design of a special machine is a highly personal project, requiring the insight, talent and dedication of a single designer from concept to installed operation on the customer's floor. When the design function is departmentalized and depersonalized for the sake of "efficiency," I believe that the inner consistency of the machine is compromised to the detriment of machine function, quality, and reliability. Even small design-and-build houses experience the trauma of a machine which gets bounced from one designer to another. It rarely works up to par, it always costs more than anticipated, and it invariably loses money for the company.

For the same reasons that large design and build companies have problems with special machinery, so do many manufacturing firms who attempt to do their own design and build in-house. In this case, the bureaucracy is already in place, denying the proper environment for top-flight machine design and build. Here we run into the machine which is truly designed by committee. There must be cooperation (seldom obtained) between Production, Manufacturing Engineering, Quality Control, Engineering, and Purchasing, among others. Total responsibility for the machine, from concept to completion, is not readily given to one person, and if it is, it is not usually the individual who actually does the work. Many manufacturing firms, after trying once or twice to design their own special machinery, often seek the independent specialist to do the job in the future. Those firms which are successful at in-house design and build of special machinery are invariably those which provide the proper environment, and more often than not, that environment turns out to be very similar to that seen in small independent design-and-build shops.

The discussion of the design-and-build business which follows takes the point of view of the small independent machine builder, although, for the reasons presented above, this viewpoint is relevant to successful larger machine design and build specialists, and to specialist groups within manufacturing firms.

3.2.   THE DESIGN-AND-BUILD PROJECT

Identification of customers who require special machinery is a marketing function which is carried out in numerous different ways. The definition of the customer's need likewise spans a wide range of detail, from the verbal statement, "We need an automatic machine," to a voluminous, detailed Request for Quotation (RFQ). It is extremely important that the design engineer who will be in charge of the machine, given that the firm gets the job, be brought into the project as soon as the potential machine is identified.

Fig. 3-1 depicts the typical conduct of a design-and-build project from the basic concepting of the machine through installation and proof-of-

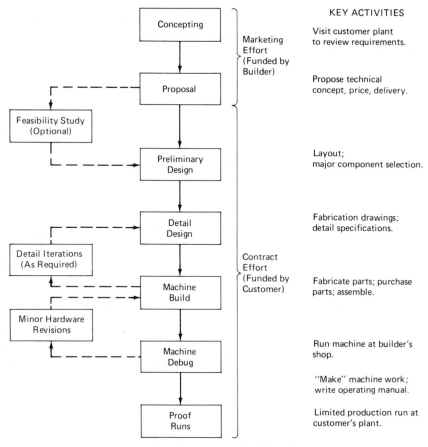

Fig. 3-1. The design and build project.

function in the customer's plant. Note that all the stages are technical in nature, requiring the participation of the design engineer. It is the author's strong opinion that a top-flight machine is one which is followed from start to finish by a single individual who conceives the basic machine, personally lays the overall machine out (or oversees the layout very closely), has the last word in component selection, closely follows the fabrication and assembly process, and lives (and dies) through debug and proof of function. There are very subtle insights and reasons for every detail which are not always fully communicated to others by the designer. These insights are lost if the project is turned over in midstream to another project manager. Similarly, the subtle essence and internal consistency of the machine is often lost if it is designed and built by a departmentalized bureaucracy.

The sections which follow describe the various stages of the typical design-and-build project in some detail. As a starting point, it is assumed that a customer requiring a special machine has been identified, and has invited the prospective designer over to discuss a potential design-and-build project.

3.2.1. Concepting the Machine

Concepting is a process which requires a broad general knowledge of manufacturing techniques, a good understanding of the customer's specific requirements, a very broad knowledge of machinery and machine components which do similar jobs, creativity, imagination, and a bit of showmanship.

More often than not, the company which eventually gets the job is the one whose cognizant designer leaves the best impression with the customer's manufacturing engineers during the first, only, and often brief visit to the customer's plant. During this visit, the designer is hearing the specific requirements for the first time and must do some fast thinking in order to give some indication to the customer as to how the actual machine might be made. Manufacturing engineers are themselves experienced in automatic machinery, particularly those common in their own industry, and have some idea of how the machine should be designed themselves. In short, it is very difficult to impress knowledgeable engineers on their home floor about designing a machine to which they have given much more thought than you have.

Under these conditions, perhaps the greatest asset to the designer is his breadth of knowledge concerning alternative ways to do the job, his ability to offer several innovative concepts which may never have occurred to the customer. It is also very important that the designer listen to and completely understand the customer's requirements. The discussion of a specific design method for concepting machinery in Chapter 4 lists a number of specific questions which the designer must try to answer before committing to a machine concept.

Concepting the machine is basically a process of generating possible alternatives, then selecting the one which best meets the customer's needs. Obviously this is easier said than done, since experience and creative insight play a large part in the generation of good alternatives.

"Quick and dirty" modelling is often used to actually see if an idea works. It can be as simple as seeing if the product will slide on a particular way design, or as complex as fabricating a cardboard mock-up to see if various parts orienting or transfer schemes will actually work. It is good practice to get a sample of the customer's product upon which the machine will operate. Simply holding the product in his hand will give the experienced designer insight into how it may be held and what existing machine components might be used to transfer, clamp, or operate on the product, as well as a general feel for any special handling precautions which should be considered.

The actual output of the concepting phase will often be a single drawing or sketch, preferably isometric (3-D) which shows the general material transfer scheme (rotary, in-line, carousel), the work stations, and the major machine components (drive motors, parts feeders, etc.). It is usually sufficient to describe the control system in words, since a drawing would not really fully describe the control concept. It is essential to show or to describe in words how all critical or exotic processes will be accomplished. The customer must be convinced that the machine will actually work, and it takes only one poorly designed aspect to render the machine an expensive decoration.

Often it is not possible to concept a machine with much assuredness that it is going to work unless a great deal of experimentation or preliminary development work is first done. In this case, a customer-funded feasibility study (Section 3.2.3) may be in order to further investigate the feasibility of the concept.

3.2.2. The Design-and-Build Proposal

The proposal is an offer to the customer by the builder to provide a machine to perform the required job. Its basic components include:

- Concept drawing or drawings describing just "what" is proposed to be delivered;
- Machine specifications describing just what the machine will do;
- A schedule by which the machine will be designed, built, and delivered to the customer's plant;
- How much the machine will cost the customer, and the terms of payment.

A proposal cannot be made without a concept to propose, although it is not uncommon for the customer to issue a Request for Proposal (RFP) which leaves very little creative latitude to the designer. In other words, the customer often tells the designer what the concept will be and requests price and delivery of the machine.

It is important to note that both the concepting and proposal stages of a design project are almost always funded by the design firm, not the customer. (For very large jobs, manufacturers have been known to fund proposals, mainly to assure that an excellent concept is proposed.) As a result, concepting and proposals must be done quickly and efficiently, well enough thought out to obtain the job, but not at a cost far exceeding the expected value of the potential job.

3.2.3. The Feasibility Study

Often a firm needs a machine, but are not sure of exactly what they want because similar machines have never been built before. They are not sure, for example, whether or not a certain process or operation can be automated, and if so, how it would be done and at what cost. If the machine requirements are particularly complicated, that firm may agree to pay a design-and-build house (in which they have confidence) to perform a feasibility study to investigate various concept alternatives in some depth and to propose and cost a reasonable machine design.

The feasibility study itself is an in-depth machine concepting project following the steps described in the design concepting process of Section 4.3. The advantage of a customer-funded feasibility study is that the design firm can afford to spend considerable time understanding the problem, considering new, creative approaches to solving the problem, and even fabricating models of critical parts of the machine to make sure that the proposed concepts will work. Funded feasibility studies are always welcomed by the design-and-build house because it indicates, first, that the customer is really serious about building the machine, and second, that they have the inside track on obtaining the full design-and-build job.

3.2.4. The Preliminary Design

The first step in the design-and-build process after the customer and the designer have entered into a contract to build a machine is the preliminary design or layout phase. This consists of preparing a full- or part-scale drawing of the entire machine, showing the locations of all work stations and the configuration of the mainline and crossline material transfer system. All active purchased parts are identified by manufacturer and model number (i.e., motors, transmissions, cylinders, valves, bowl feeders, pick and place units, actuators, work heads, rotary table, etc.). The control system is specified in terms of the control scheme, control hardware, and if programmable, a block diagram of the software.

One might ask, "How can one quote a job accurately without having already identified at least the purchased hardware, its source, and its cost?" The answer, of course, is that one cannot; therefore, much of the preliminary design must be done during the proposal stage. The designer does learn from experience what the large-cost items in a particular type of machine tend to be, and can estimate costs fairly accurately without costing each item exactly. Obviously, the designer must be adept at cost estimating, as the design firm stands to make or lose money as a result of the machine cost estimate.

Many design and build projects include a design review after the preliminary design phase has been completed. Before the designer expends great amounts of time preparing detail drawings, it is wise to verify that the customer understands fully what he is getting for a final machine. Often the preliminary design presents a different perspective as to how the machine will actually look than did the concept drawing prepared for the proposal. If the customer has any inputs at this time, it is important to get them while changes can be made with relative ease.

3.2.5. The Detail Design

In a design-and-build project, detail design does not mean that the designer is finally considering the details of how the machine will work. Rather, he is finally considering the details of how the machine will be built, and in the process providing the builder (i.e., the shop) with detail drawings from which machine parts can be fabricated.

Machine details can generally be released to the shop for fabrication as they are drawn. This is a further indication of the fact that the machine has already been designed, even to the last detail. It cannot be overemphasized that functional details must be considered as early as the concepting stage—they cannot be considered after the machine is actually in construction.

One very important aspect of machine detailing is tolerancing of components. Preliminary design, or even concepting, assumes that surface A can be made perpendicular to surface B to 0.0001 inch. The detail design determines that intermediate machine members must be fabricated to dimensional and/ or form tolerances of 0.000025 inch (or whatever) in order to result in the desired machine configuration.

3.2.6.   Machine Build

The machine is built from detail drawings generated in the detail design phase. Unforeseen problems always occur concerning clearances, access to certain areas, interferences which were not anticipated, etc. Thus a certain amount of iteration takes place on the detail design of various components as the machine is being built.

Since most automatic machines consist of a large number of purchased components, the build phase also includes purchasing, expediting, and inspecting purchase parts. Delivery times for purchase parts are usually the schedule limiting factors, so the machine build phase is often the longest phase, even though fabricated parts are made in relatively short periods.

There are two methods of building special machines, the choice of which is determined by factors beyond program management. The first method consists of waiting until all fabricated parts and purchased parts are in house before starting machine assembly. The machine is then built up with a concentrated effort by technicians, support machinists, control engineers and electricians. This method has the advantage of the dedicated efforts of those involved and minimum tie-up of floor space. The greatest disadvantage of this method is that assembly problems all show up at the same time. The specialists assigned to the machine end up spending more time with the machine, much of it unproductive because they are waiting for problems to be solved.

The second method consists of scheduling fabricated parts so that the machine base or chassis is made first, and various work station components made later. The machine can thus be built up as parts are made or received from vendors. This method ties up assembly floor space and requires a good deal of project planning and management. It offers the advantage of flexible use of support personnel and identification and solution of problems on something less than a crisis basis.

Obviously, the method used by a specific builder will depend upon how much floor space he has, what percentage of the machine he fabricates in house rather than subcontracting it out, and the number and type of support personnel he has available. In practice, some intermediate method of machine build operations will evolve.

3.2.7.   Machine Setup, Debugging, and Proof

The final phase of machine design and build takes place at two locations—the builder's shop and the customer's production floor. It is a rare occasion indeed when the machine operates to specification immediately after assembly. Making the machine perform to specification, both at the builder's location and on the customer's floor, is the process known as _debugging_. Even with a well designed machine, this stage can consume up to 25% of the total machine cost.

Each independent subsystem is first cycled slowly and independently of the other machine subsystems. In general, this means assuring that the material transfer system moves typical workpieces from station to station with maximum positional accuracy and repeatability. The crossline material transfer system, or parts feeding lines, are tuned so as to keep the machine supplied with parts, even if the machine were running at design speed. Work stations are tested to assure that their operational function is performed satisfactorily and reliably for each actuation control input signal. Having determined that each machine subsystem operates perfectly and repeatably by itself, the next step is to run them in unison.

The entire machine is first cycled slowly to insure that synchronization of the subsystems is present. Machine speed is then increased until problems occur or until design speed is reached without problems. Each auxiliary control function, such as safety interlocks, automatic shutdown due to sensed irregularities, automatic parts rejection, etc., is checked by simulating a malfunction. Although it is impossible to anticipate every possible malfunction, as many as can be conceived of should be tested. If the basic control function and the auxiliary functions check out at slow machine cycle speed, then the control algorithm is fundamentally sound. Many problems do arise, however, as machine speed is increased. Generally, the control system is not acting and reacting fast enough to handle operation at high speed. If this is the case, the control algorithm must be revised, or implemented in a faster way (e.g., solid state logic instead of relay logic, electro-pneumatic logic instead of pneumatic logic, control program written in machine language rather than in BASIC, etc.).

Another problem which occurs at high-speed cycling is synchronization offset. The machine may have worked perfectly at low cycle speeds because the synchronization error was a small, and thus an inconsequential percentage of the total machine cycle. At high speed, that error is still small in absolute terms, but large compared to the machine cycle. To correct this type of error, the control signal lag must be shortened, or if this is not possible, the guilty device must be made to anticipate its actuation signal, or the actuation signal itself advanced in time.

Finally, the problem of workpiece inertia comes into play at high machine speeds. Very light pieces which are intermittently started and stopped tend to slide or fly away, therefore must be contained. Heavy work which is rapidly accelerated and decelerated can deflect light machine members. Intermittent starting and stopping provides an excitation for machine vibration which usually can not be avoided. The solution is to beef up critical machine members so that they have resonances outside the range of excitation, or so that displacements are small enough not to affect machine operation.

Debugging and adjusting high-speed machinery is a difficult job, and should always be done in a systematic manner, following the steps outlined above.

Once the machine has been debugged at the builder's shop, it is usually demonstrated to the customer by cycling the machine at design speed using a limited number of actual workpieces supplied by the customer. Then the machine is partially disassembled and transported to the customer's shop where it is reassembled and the debugging procedure repeated. Debugging in the customer's plant usually only requires adjustment, as opposed to major hardware changes or modifications. Final proof of the machine consists of demonstrating a run of several thousand pieces through the machine at design speed with fewer than a specified number of rejects and/or machine jam-ups. Often a significant portion (usually $\frac{1}{3}$) of the machine price is withheld by the customer until this final proof of machine function at the customer's plant.

4. THE DESIGN PROCESS

4.1. DESIGN AND CREATIVITY—A GENERAL OVERVIEW

There is, I believe, the scientific equivalent of an artist in every design engineer. At one time or another we have all experienced the fantasy of sitting down to the drawing board and creating the perfect engineering entity—whether it be a machine, a bridge, an instrument, a circuit, or whatever. We envision "perfection" as harmony of function, optimum performance, infinite operational life, elegant simplicity, and pleasing aesthetics. We strive to be "creative," in that the object of our effort will be unique (nothing like it has ever been seen before); clever (our peers will marvel at the obvious now that we have pointed it out to them); and historic (our device will be referred to in future texts as an example of creative design).

Obviously, we rarely, if ever, fulfill this fantasy. In fact, we rarely get the opportunity to design in a completely freewheeling manner without practical constraint. How often do we hear: "Make it fast," "Make it cheap," "Make it out of material we already have," "Copy the competition," "Make it last until next Christmas," "Do it the same way as you did the last job," "Do it my way," "Don't worry about how it looks?" These constraints are not only real—in most cases they are entirely legitimate. There is, however, room for creativity in engineering design, regardless of constraints.

There is not, unfortunately, a universal design process which offers a step-by-step procedure or checklist for the creative design of engineering objects. There are, however, numerous design methods which apply very well to some specific design problems. Jones[1] describes over 35 different design methods and techniques in some depth, demonstrating the large number of formal design methodologies which have evolved over the years. Table 4-1 presents some of the better known methods and briefly describes their objectives.

Engineering design is inherently a creative endeavor, in that all designs attempt to assemble known components into a new or different arrangement

[1]*Design Methods.* J. Christopher Jones Wiley Interscience, London, N.Y., 1970.

TABLE 4-1. Design Methods.

Method	Objective	Features
Brainstorming	Generate many ideas in a short period of time without regard for practicality.	Group effort to come up with as many creative design ideas as possible. No in-depth discussion of ideas. No constraints on how "far out" ideas can be. No criticism of ideas allowed.
Synectics	Group effort to work several creative alternatives to their practical implementation.	Tightly chaired group effort by "experts" to transform one or more creative ideas into workable designs. Discussion guided to obtain consensus conclusions, not just ideas.
Morphological charts	Assure that a potentially viable design approach is not overlooked by considering all possible alternatives.	Make highly structured, n-dimensional matrices listing all possible design parameters for each of the independent design variables. Methodically eliminate unfeasible combinations, develop feasible combinations further.
Ideation	Generate creative ideas for designs or design improvements through association with possible changes.	Question design by asking how it could be: improved, modified, rearranged, reversed, adapted to new uses, made larger, made smaller, combine functions, made more convenient, safer, lighter, etc.
Functional visualization	Avoid starting a design with preconceived solutions by proper statement of the problem.	State the problem in its most basic functional terms. Define design object in terms of what it must ultimately do, not what it must be like; e.g., "Design a method of making metal cylinders," not "Design a lathe."
Value engineering	Make object cheaper without losing overall functionality.	List all cost-producing elements. Question their need as regards function. Eliminate nonfunctional elements. Combine functional elements. Scrutinize processes.

to solve a specific engineering problem. If the result is in fact new, then the design is a "creation." If the result is identical to previous work, it is a copy—no matter how well intentioned the effort may have been.

A great deal has been written about creative design and those that possess the talent. One oft-reached conclusion is that creativity cannot be taught, but that it is inherent in the individual. I would argue that the lightbulb flashes above one's head (referred to in the literature as "illumination" or "inspiration") without regard for inherent creative talent, but rather as the result of a particular visualization of the problem coming into mental proximity to a particularly good, concise solution at the same instant. This being so, then creative inspiration is a chance occurrence, but the probability of that occurrence can be enhanced by seeing that the problem is clearly understood, and that a large number of potential solutions are considered. Thus the probability of generating truly creative designs would be significantly greater for those individuals who are able to clearly understand and concisely define the design problem, and who have a large reservoir of potential solutions stored in experience and/or general knowledge. The probability can be further increased by developing an efficient technique for bringing the problem into close mental proximity to the solution(s) so that "inspiration" can indeed occur.

If creativity, then, can not be taught or cultivated, certainly the contributing circumstances can. This is the basic assumption behind the design method presented in Section 4.3. With respect to engineering design in general, and the design of automatic machinery in particular, the following assumptions are made:

1. A designer can be taught to identify and define design requirements in such a way that potential solutions will not be overlooked. With respect to special machinery design, this includes close study of the process being automated, and asking of appropriate questions in search of what is *really* required.

2. A designer can be taught (often referred to as *passive experience*) or can learn by doing (*active experience*) specific design solutions to various engineering problems. In design of machinery, useful experience includes extensive knowledge of components available to meet various machine requirements, and specific applications which have worked under different circumstances.

3. A designer can be taught a *design method* or *design approach* which forces him to systematically consider alternative solutions to the design problem, and to evaluate the alternatives in an intelligent manner, leading to the best (optimum) solution under the given circumstances.

Special machinery design is one of the few areas of design engineering which offers a high degree of independence, creative latitude, and total project involvement to the designer. The environment is one in which every job is different and, to a certain extent, requires a novel, if not creative approach to the problem at hand. As pointed out in Chapter 3, the general business environment and project conduct requires that the designer be not only creative, but prolific in his ability to concept machines. For every machine actually designed and built, probably five excellent design concepts have been proposed.

4.2. SYSTEM DESIGN

Fig. 4-1 depicts the essence of system design. It is the combination of single-function elements in a way that renders some well defined set of inputs transformed into a desired output. There are system constraints which limit design freedom and which impose additional requirements on the design. The system design problem is one of combining elements into a complete system which yields the desired output(s) for given input(s) while simultaneously satisfying all system constraints. Since there is usually more than one acceptable solution to the design problem, the question arises as to the best, or *optimal* system design.

The fact that there is always more than one acceptable solution to the automatic machine design problem is, I believe, the reason that many engineers feel threatened by the task of designing a complete machine. It takes a great deal of confidence in one's abilities to propose a machine which is different from others of similar function, particularly if a previous design has worked satisfactorily. The key question, however, is whether or not the machine (system) will give the desired outputs for the given inputs while meeting all other imposed constraints or requirements.

In designing the special production machine, the desired output is the finished workpiece, whether it be a machined part, an assembly, a packaged item, or simply data from a measurement. There are certain quality specifications associated with the finished workpiece which define an acceptable output. The machine input or inputs may be raw material, partial assemblies, finished assemblies for inspection or packaging, or parts to be added to the main workpiece. As does the output, so does the input have quality specifications associated with it, and it is imperative that the designer accurately assess the complete nature of the system (machine) input(s) before designing the system. System constraints include machine cost, overall size and weight, production speed, maximum reject rate, environmental interface requirements, level of operator skill, and specific customer requests. If the

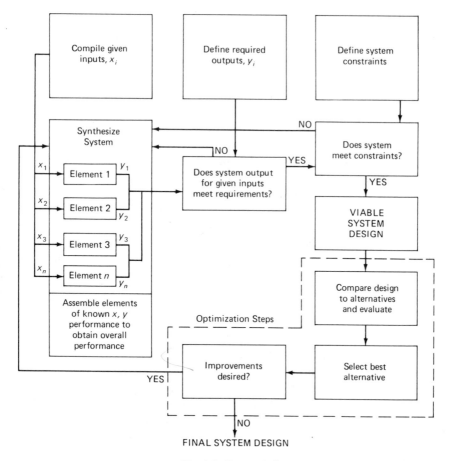

Fig. 4-1. System design.

customer says that the machine shall be painted blue, then that is a design constraint.

There are machine elements, both off the shelf and designed from scratch, available to the designer to perform small steps of the total system transformation from input to output. Some perform their individual functions more efficiently than others, yet the others may offer another advantage, say accuracy of operation. They all have different price tags. It is the job of the machine designer to consider all the options available, consider the trade-offs and relative importance of each, then specify a system design which satisfies all constraints. All design decisions must be internally consistent if the design is to be optimized. For example, one would not fixture a top-quality indexing head with inferior tooling just because it saves cost.

The question of whether a particular machine design is indeed optimal is often subjective and dependent on the observer's assessment of the relative importance of design trade-offs. A pneumatically driven machine may meet all design requirements and yet be less expensive than a similar design which is electrically driven. However, the customer may not share the designer's opinion that this is indeed the optimum design if all other machines in his plant are electrical, and he was not particularly seeking a low-cost machine. In fact, the cost of training maintenance personnel to service the "oddball" machine may be greater than the initial saving on the machine. The next section, detailing a specific design method for automatic machinery, emphasizes the importance of understanding the relative weights which the customer places on design constraints before developing a design concept.

There is no simple, straightforward approach to complex system design. The designer must always keep the total system in perspective as each of the system components is selected.

4.3. A DESIGN METHOD FOR AUTOMATIC MACHINERY

The design method described in following sections has been developed so as to be as general and universally applicable as possible. Obviously a concise, step-by-step procedure which will work every time for all machines, be they assembly machines, inspection machines, test machines, machine tools, or packaging machines can never be formulated. Therefore, some of the specific steps presented here may be applicable in some cases but not in others.

The methods presented here are practical as opposed to theoretical. The procedures are discussed within the context of systems design, but without the highly mathematical treatment which often accompanies treatises on system optimization, electrical systems, and computer systems.

The creative design work associated with automatic machine design takes place primarily during the concepting phase. This is not to say, however, that creative detail design work does not take place later on in the design process.

The design method which I propose consists of three broad phases: the *concepting* phase, the *layout* phase, and the *detailing* phase. Note that the design phases roughly correspond to the design project stages of concept, preliminary, and detail design, but do not confuse the design *project,* which is a business endeavor, with the design *process,* which is a creative endeavor.

The objective of the machine concepting phase is to select a machine system from among several feasible alternatives, which meets all input/output requirements and which satisfies all imposed system constraints. The creative effort associated with this phase is the generation or *synthesis* of feasible alternative concepts.

The objective of the layout phase is to specify and size the major functional components of the system and to determine for the first time just what the machine will look like. It should not result in a change of the basic concept, but rather serve as the transition of the machine from an idea to reality.

The objective of the detailing phase is to specify and size all machine components so that a machine can be built. Again, it should not result in a change to previous work, rather it should constitute the final refinement.

Fig. 4-2 depicts a six-step design procedure for concepting an automatic

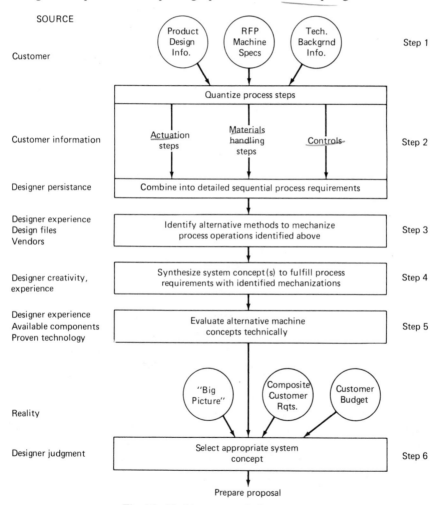

Fig. 4-2. Machine concept design process.

machine. Keep in mind that the design procedure or process presented here is an abstract framework meant to guide the designer in the formulation of the machine design. They offer an attempt at formalizing a creative effort which is usually unstructured.

The sections which follow discuss the design steps in detail.

4.3.1. The Machine Concept Design—Obtaining Required Information

The first step in concepting a machine, as it is in designing anything, is to assemble all relevant information including design requirements, background technical data, and even subjective assessments of what the customer really wants. In general, it pays to obtain all information and data before a concept is proposed. Special automated production machinery is not inexpensive. A serious customer is asking you to design and build a machine which will cost $75,000 and up, and which will, through its operation, contribute product sales and profits on the order of millions of dollars. It is not in the customer's best interest to withhold information, no matter how insignificant it may seem. The machine designer therefore usually finds the potential customer very willing to supply information and clarify needs.

Basic information required to develop a machine concept includes the following:

- Machine specifications and required performance;
- Drawings of product assemblies and components which the machine will have to handle or otherwise interface;
- Several sets of actual piece parts, if available;
- A concise definition of the intended machine/environment interface.

If the product is already in production, even manually: (a) request a plant visit to view the operation; (b) request operations sheets, assembly diagrams, or any other current production documentation.

Table 4-2 provides a list of detailed technical information requirements. If this information is not provided with the customer's RFP (Request for Proposal) it should be requested. A well prepared RFP will include all information required to make a viable proposal. A poor RFP may only include a vague description of what the machine will be required to do, and should be accepted with caution. If the potential customer is not sure himself what he wants to do, there is a good chance that he is not yet serious about purchasing a machine. In this case, any information supplied by the builder is simply free consultation being used to educate the customer.

TABLE 4-2. Technical Information Requirements for Automatic Machine Design.

1. Desired production rate
   A. Gross rate over specified time *A − B = C (?)*
   B. Net rate over specified time
   C. Reject rate
   D. Machine efficiency (derived from A and B) *B/A*
2. Functions to be performed (if specified) or, if left up to designer; required production steps, including:
   A. Number of steps in process
   B. Number of parts
   C. Optional sequences vs. fixed sequences
   D. Subassembly inspection between steps
   E. Allowable manual intervention
3. Machine/environment interface
   A. Input interface
      1. Type of loading   *e.g. in line with*
      2. Details of input machine interface (if required)   *existing equipment*
   B. Output interface
      1. Type of unloading
      2. Output machine interface (if required)
   C. Parts loading details *(unloading)*
      1. Bulk
      2. Magazine
      3. Manual
      4. Other   *(blister)*
   D. Human interfacing with controls
      1. Alarms
      2. Visual data or status readouts
      3. Manual overrides
   E. Auxiliary services interface
      1. Electrical (voltage, phases, service)
      2. Air (pressure, capacity)
      3. Water (pressure, flow), e.g., for cooling
      4. Waste disposal (as required by process)
      5. Gas (for heating, drying, heat treating, etc.)
      6. Vacuum
   F. Physical environment
      1. Temperature
      2. Humidity
      3. General cleanliness
   G. Machine isolation for safety (guards, screens, etc.)
   H. Responsibility for all of above: Which items must the customer supply and which must the builder supply?
4. Skill level of operators, users
   A. Skilled technicians
   B. Unskilled factory labor
   C. Malicious idiots

TABLE 4-2. (*Continued*)

5. Product characteristics, properties, limitations
   A. Sacred surfaces
   B. Strength under loading in various directions
   C. Forbidden substances
   D. Unique properties of customer specified parts, materials
   E. Concise definition of acceptable product (or, conversely, of a reject)
6. Past experience with product
   A. How do actual parts differ from drawings?
   B. In the case of parts, what is typical percentage of bad parts (i.e., capable of causing a machine stoppage)?
   C. What production steps have presented problems in the past?
   D. Identify problems—causes if known.
7. Product details
   A. Assembly drawings
   B. Subassembly drawings
   C. Component drawings (i.e., those to be fed or otherwise interfaced with the machine)
   D. Quality requirements
   E. Parts list
   F. Vendor list (for parts to be used with machine)
8. Customer-unique specifications
   A. In-house standards (wiring, safety, etc.)
   B. General workmanship requirements
   C. Finishes (corrosion protection, paints, colors, etc.)
   D. Other

4.3.2. The Machine Concept Design—Quantizing Process Requirements

The second step of the machine concepting process (Fig. 4-2) is one which is unfortunately given too little or no attention by the designer under pressure to concept a specific machine. Too often the designer jumps immediately into the mechanization of a process without concisely defining the individual process steps independently of the hardware required to automate the process. One possible result of this common oversight is that the designer later compromises overall machine performance or product quality so as not to have to make a significant change to a hastily conceived machine concept which "seemed appropriate at the time" and has now progressed quite far toward completion.

Defining process or machine requirements is not a particularly creative or exciting job, which explains why it is often given only minimal consideration by the designer. It first requires understanding exactly what the customer wants the machine to do. It is easy to assume that you understand what is required after the first visit to the customer's plant, but after considering just what the machine must do, subsequent interaction is often necessary. This task also demands consideration of every process detail, even those which may seem minor but which nevertheless must be performed, sometimes to the extent of requiring a separate work station. Table 4-3 indicates some of the minor but necessary process considerations which may in fact be harder to mechanize and automate than the major process step itself. If the product for which the machine is being designed is already in production, then these minor steps will be known to the customer, and he should have specified these details in his RFP. If the product has not yet been produced, then the customer himself may have overlooked the need for some of the more subtle process requirements.

Overlooking a minor process requirement can be rectified relatively easily if the mistake is recognized before a proposal is made. If the mistake is uncovered later, it can be fixed during the design phase of the project without

TABLE 4-3 Frequently Overlooked Process Details

Major Operation or Process	Complementary Process Details Requiring Design Consideration
Machining	Deburring
	Chip removal
	Coolant application
	Tool sharpening or changing
	Wheel dressing
Assembly	Precise work positioning
	Parts feeding
	In-process inspection
Inspection	Calibrate sensor(s)
	Zero reference
	Clean work to assure accurate reference location
	Clean work holding fixture
Painting	Clean surfaces
	Prime surface
	Cure or dry paint
	Solvent removal (recycle or disposal)
Adhesive	Surface preparation
	Curing the adhesive
	Keeping application devices clean
Soldering	Clean surface
	Fluxing

much difficulty, but the machine may now be somewhat different from what was shown in the proposal. Who pays the extra cost of this modification? Obviously the answer depends on many factors which have nothing to do with the design of the machine. Needless to say, it is embarrassing to the builder who appears to have forgotten that a machining operation raises a burr which must be removed—regardless of the fact that the customer's RFP did not specifically call out a deburring operation.

Sometimes the machine can be completely assembled before it is discovered that additional processing steps should have been included. Usually it is possible to add on sufficient mechanism to solve the problem, and many such after-the-fact fixes are indeed quite clever and well engineered. Unfortunately, they still look like add-ons and do not speak well for the machine builder who is expected to design the machine right the first time. The fact that the customer was negligent in his original specifications is quickly forgotten. That "klugey looking machine" stands as a monument to the builder for many years to come.

If the machine is being designed to automate a process currently in operation, say manually, then the existing production operations sheets, if they exist, are invaluable. "Op-sheets" break a process down into its most basic sub-operations, often including steps which tell the worker where to put his hand to pick up the next part. Operations sheets are designed to instruct even the dumbest worker how to perform an acceptable job. But keep in mind that the machine is dumber than even the dumbest worker. It is therefore imperative that the process be defined completely and to the most basic detail.

There are two physical activities performed on the workpiece as it is processed by a machine. The first includes all actual operations performed on the workpiece by the machine work station(s) or machine actuators. The second includes transport of the workpiece through the machine as operations are performed on it. It is therefore convenient to consider machine requirements in terms of the generalized automatic machine model of Section 2.3:

1. *Work station requirements:* What operations must the machine perform on the workpiece so as to transform the input material into the desired output, or final product?
2. *Material transfer subsystem requirements:* What motions, including changes in orientation, must the workpiece go through as the various required operations are performed?

The third generalized machine subsystem, the controls, do not physically interact with the workpiece, and thus must be considered not as process requirements, but as the means of effecting work station and material transfer operations.

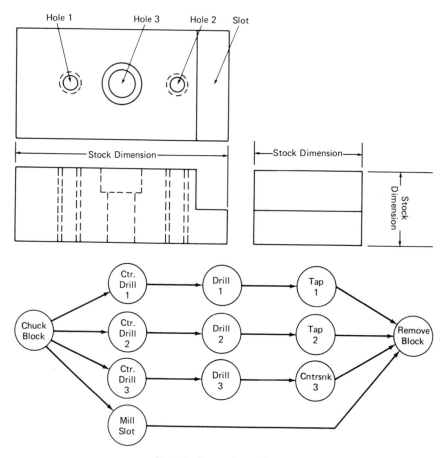

Fig. 4-3. A precedence diagram.

When a number of operations are to be performed on a workpiece, and there are several options as to the order of process operations, it is often useful to develop a precedence diagram which shows graphically all of the possible operations sequences. Fig. 4-3 shows a sample precedence diagram.

For each possible sequence of operations derived from the precedence diagram, there is a fairly specific, time-phased sequence of process operations and material transfer. Clearly, designation of specific transfer needs presumes certain operations or combination of operations taking place at one machine location (i.e., work station), and this in turn presumes a certain type of work station actuator. Recall that this step in the machine concepting process is to ascertain *what* must be done to the workpiece without concern for *how* it is to be done. A certain degree of practical judgment must be exercised, however, in generalizing what is required by the machine.

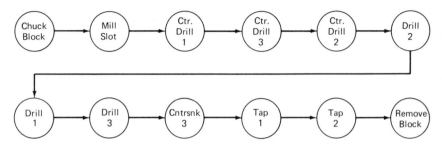

1. Load block in fixture on x-y table
2. Clamp block
3. Table index to miller station
4. Miller on
5. Cutting fluid on
6. Table traverse work to cut slot
7. Cutting fluid off
8. Table index to drill station (turret) hole #1
9. Turret index to center drill
10. Center drill on
11. Feed center drill (or work)
12. Retract
13. Index table to hole #3
14. Feed center drill
15. Retract
16. Index table to hole #2
17. Feed center drill
18. Retract
19. Center drill off
20. Index turret to drill for hole #1, #2
21. Drill on
22. Coolant on
23. Feed drill, drill #2
24. Retract drill
25. Index table to hole #1

26. Feed drill, drill #1
27. Retract drill
28. Index turret to drill for hole #3
29. Index table to hole #3
30. Feed drill, drill #3
31. Retract
32. Index turret to countersink
33. Feed countersink
34. Retract
35. Coolant off
36. Drills off
37. Index table to tap station, hole #1
38. Tap on
39. Tapping fluid on
40. Feed tap
41. Retract
42. Index table to hole #2
43. Feed tap
44. Retract
45. Tap off
46. Fluid off
47. Index table to unload position
48. Release clamp
49. Unload work

Fig. 4-4. Process requirement list.

The most practical graphical device for specifying process requirements is the simple list. Fig. 4-4 shows such a list which not only outlines process requirements, but also shows material transfer requirements in the proper sequential perspective. Note that there may be several lists possible, one for each allowable sequence as determined from the precedence diagram. If the designer desires to be more graphical (e.g., for a preliminary presentation) the list can be drawn as a block diagram. Block diagrams have the advantage of being able to show sequences of crossline operations and materials transfer in the case of multiple interacting processes. The more elaborate the quantization of machine requirements becomes, the longer this step in the concepting process takes. The essential requirement, however, is a detailed list of required operations.

Before progressing to the next design step, it is appropriate to point out one conceptual aid to machine requirement quantification. Try to visualize the workpiece moving in space from machine input to machine output *but with no machine present*. Visualize the product as each operation is performed and the manner in which the product must be moved and/or oriented in order for each operation to take place. This thought exercise not only helps to clarify machine requirements, but it also aids in the later step of synthesizing a machine configuration within which this imaginary sequence of product operations and motions take place.

4.3.3. The Machine Concept Design—Identifying Alternative Mechanizations   *incl. controls*

There are many designers who believe that "tried and true is better than new" and, as a result, will concentrate on copying previous, similar machines to meet a particular requirement. Their concept design process thus becomes very straightforward, and in most cases very efficient. The complete machine is modelled after a previous machine which served a similar purpose, and all of the details are bent to fit that model. It is difficult to fault this technique, particularly when the final machine works and works well.

Rarely, however, does the special machine designer run across *exactly* the same machine requirement as encountered in the past. Often the overall problem is similar to past machine requirements, and often elements of the overall problem are in fact exactly the same. However, in my opinion, each machine design problem warrants consideration of more than just one practical solution to that problem. There is nothing at all wrong with modelling the first alternative after a previous machine, and upon further consideration, that first alternative may well prove to be the best solution.

The design method presented in Fig. 4-2 facilitates consideration of alternative mechanizations of the process requirements simply because these requirements have been listed in detail. It is a relatively simple matter to add to the list of Fig. 4-4 various alternative devices for performing each process operation (i.e., ways of implementing each work station) as well as alternative ways of effecting workpiece or parts transfer. This step, like the one before, does not require a great deal of creativity, but designer experience and awareness of what off-the-shelf devices are available is very important.

To this point, very little in the way of designing has taken place. The designer has simply taken great pains to understand what is required of the machine, break these requirements down into their most basic detail and then identity various ways to actually perfrom each basic operation. It should be noted that the three steps of the concepting process discussed so far *need not take a great deal of time* to execute. In fact, the well prepared designer may well finish the operational breakdown during his initial visit to the customer's plant, and have identified several ways to mechanize each just from his experience and background knowledge. It would not be a bad idea to review the detailed list of process requirements with the customer's production engineer(s) before leaving the plant just to be sure that a critical step has not been overlooked.

Step 2 considered only the machine requirements from the work station and the material transfer points of view because the workpiece itself really does not have any inherent machine control requirements. However, specific mechanizations of work actuation and work transfer *do* have specific control requirements associated with them, and Step 3 of the design concepting process should consider these requirements. The control system factors of interest at this stage of the design concept are both power and control signal inputs to the work stations or work transfer devices under consideration. Practically speaking, this means additional columns in the list of Fig. 4-4 to specify power and control requirements for each process step mechanization under consideration.

Many machine designers, having primarily mechanical backgrounds, question the need for considering the control system at this early stage of machine concepting. They argue, and rightly so, that a control system can be designed to interface with any eventual combination of work stations and work transfer designs. But consider the complexity and interfacing problems which would occur if the machine components and devices selected required power inputs including 2500 psi hydraulic oil, 80 psi air pressure, 440 volts AC, single-phase 220 volts AC, three-phase 220 volts AC, and control inputs including 15 psi air pressure, 28 volts DC, and 110 volts AC. This type of a component selection, while exaggerated, is made even worse if it turns out that the same result could have been achieved by selecting components requiring only three-phase 220 volts AC as a power source and 110 volts AC as control signals.

The type of problem described above is more prevalent in the design of automatic inspection or automatic test machinery which requires exotic electronic equipment as part of the machine function. In some cases there may be no choice but to include the design of a complex interfacing subsystem into the machine because the required component is the only alternative available, and the designer must use it as it stands.

The result of Step 3 in the concepting process need be no more than another list or additional columns in the Step 2 list of Fig. 4-6 which contain alternative components or custom designed mechanisms which will do the required job. It is also useful to include power and control data as mentioned

1 obtaining Req. info
2 quantitizing process Req.
3 Identifying alternatives mech.

above. It is possible to add alternative actuating devices to a block diagram of the detailed process requirements, but the chart becomes quickly crowded.

4.3.4. The Machine Concept Design—Synthesis of System Concept(s)

The combination of system elements in such as way that the resultant system transforms inputs into desired outputs while satisfying all system constraints is known as *synthesis*. Synthesis is the truly creative aspect of system design, or in our case, machinery design.

Steps 1–3 of the concept design process have resulted in identifying the system elements with which we have to work. The detailed breakdown of required operations has pretty much defined the order in which these elements must be arranged (although there may be some alternatives). In its simplest form, the design or synthesis of the machine requires the mounting of these identified elements onto a table in the proper order. This design may meet the input/output criteria of the system design, but will most likely fail to meet one or more of the "system constraints" which apply to the machine.

Table 4-4 lists system design constraints which usually must be met in designing a machine. These constraints are listed roughly in order of

TABLE 4-4. Concept Design Constraints.

Design Constraint	Source of Constraint	Design Implementation
1. Production rate	Customer specification	Select work stations, transfer systems which will cycle at required rates. Analyze probabilities of finished part rejection to realistically determine machine efficiency. Design cycle rate at least 25% higher than net required production rate.
2. Tendency to foul (reject rate)	Customer spec.; quality of parts; nature of process	Analyze potential reject or failure modes and provide means for avoiding. (Inspect parts on line, presort parts, detect worn tools, etc.)
3. Ease of unjamming (time to clear)	Designer	Design for easy access to workheads, independently movable tools, etc. All potential hang-up points should be accessible to operator without tools (spring latches, $\frac{1}{4}$ turn fasteners, toggle clamps).
4. Accuracy	Product quality requirement	Always design machine to tolerances one order of magnitude finer than required product tolerances.

TABLE 4-4. *(Continued)*

Design Constraint	Source of Constraint	Design Implementation
5. Reliability	Customer spec.; company reputation.	Anticipate imbalances, vibrations, dynamic stresses, stress concentrations, fatigue failures, and design to avoid. Keep all electromechanical devices (motors, solenoids) cool (below $130°C$). Avoid side loads on cylinder piston rods.
6. Safety	Product liability	Prevent operator's hands from entering workhead active space by guards. Provide automatic shutdown if guards are compromised. Provide "panic button".
7. Maintainability	Designer	Design modular machine. Allow entire subsystems to be easily removed and replaced. Allow easy access to high-failure-rate components (fuses, etc.).
8. Cost	Customer budget	Understand clearly cost constraints before starting. "Low-cost" machine is not consistent with high accuracy, reliability, and degree of automation.
9. Use of off-the-shelf components	Designer	*Always* choose off-the-shelf components over custom designed ones, if they can do the job.
10. Level of automation	Customer specifications	Always provide a manual override to allow human intervention.

importance, but the order may vary for different machine design jobs. Generally the customer defines the importance of various machine design constraints, although he is not always clear in defining their precedence. It is not uncommon for the customer to expect all factors to be considered with equal and maximum concern, failing to realize that trade-offs must always be made with some precedence hierarchy in mind.

As the designer attempts to develop one or more alternative machine concepts with the constraints clearly in mind, the synthesis task becomes more difficult and demanding of creative solutions than did the simple task of bolting work stations onto a table. The very important requirements of production rate and ability to run without jamming may eliminate some work stations or material transfer devices which were previously identified as capable of performing the required operations. Consideration of the physical size of work station alternatives may have a bearing on overall machine configuration (i.e., rotary or in-line). Certainly consideration of production rate requirements and workpiece sizes will affect decisions regarding the work transfer subsystem. The decision to use an off-the-shelf machine chassis narrows the concepts to one of in-line, rotary, or carousel chassis from one of several manufacturers. The degree-of-automation consideration will have a strong impact on the type of control system required to do the job.

In most cases, consideration of alternative component combinations in light of the appropriate machine design constraints will result in only several, perhaps only one, practical machine concept. That concept however, may still be refined even further by asking questions such as those presented in Table 4-5. Just how these questions are answered, and how those answers are incorporated in the machine design will be unique to each designer. The concepts (and final machines) which result will be a measure of that designer's creativity and talent. How the machine performs once designed and built will serve as the ultimate proof of a design's worth.

The output of this design step will be a sketch or schematic of one or more machine concepts showing the following:

- Work and/or parts flow from machine input to machine output. Material transfer configuration should show whether in-line or rotary (or other) concept is used and transfer system drive system.
- Work holding schemes. Show work carriers if required and relation to work transfer subsystem. Show method of work positioning and clamping.
- Work station location. Show relative location of all work stations mounted with respect to work transfer subsystem.
- Separate sketches of particularly intricate mechanisms which you propose to use which are not standard components. Show how each machine action will be performed.
- Word description or block diagram of control system identifying the basic logic implementation method (microprocessor, minicomputer, relay, etc.). Describe or list proposed control functions associated with machine operation.

Designers very often base the concept around a specific device, usually one with which they have had prior experience and success, which is identified by manufacturer and model number.

4.3.5. The Machine Concept Design—Concept Evaluation

Only rarely does a machine concept come together which meets all of the process requirements using the designer's first choice of work stations,

TABLE 4-5. Factors to Consider
Before Finalizing Design Concept.

Workheads or Work Stations

- Are off-the-shelf, modular work stations used at every opportunity? Are those that are used capable of required speed, power level, and accuracy:
- Can any separate work stations be combined into one?
- Is it more effective to move workheads relative to work than vice versa?
- Are workheads easily accessed for jam clearance, maintenance?
- Is there an advantage to performing operations from different directions (e.g., from sides or bottom instead of from top)?
- Is there any single operation about which you have any doubts as to speed or accuracy? If so, what are alternatives with lesser problems?
- Have reliability, maintainability, and safety been given sufficient consideration?

Material Transfer System

- Can machine be more effective (higher efficiency, faster) in another material transfer configuration (i.e., in-line, rotary, carousel, continuous, etc.)?
- Has an off-the-shelf machine chassis been considered?
- Is there a better way of actuating the proposed material transfer configuration?
- Can work carriers be used to improve work positioning accuracy? Or is transfer without carriers sufficient?
- Is the parts feeding system compatible with work transfer system (accuracy of parts position, feed rate, reliability of parts placement, reject rate, etc.)?
- Can parts feeding system be easily accessed for clearing jam-ups?
- Are there better parts feeding alternatives than those selected?

Control System

- Is total control medium internally consistent (i.e., all electric, all pneumatic, all hydraulic random mixture.)?
- Is central machine controller capable of controlling all aspects of machine operation?
- Have all possible manual inputs and overrides been included?
- Is there sufficient information transfer to operator via readouts, alarms, printouts, etc., to render most efficient operation and troubleshooting?
- Have all man/machine interfaces been considered, including clearing jam-ups, maintaining machinery, operator safety, starting and stopping machine, slow cycling (jogging) machine, etc.?
- Is sensor accuracy compatible with machine accuracy?
- Is control system response time compatible with machine cycle times?
- Have all industry standards and local ordinances been observed in machine controls (wiring, enclosures, protection and isolation of circuits, etc.)?

material transfer schemes, and control schemes, and which exhibits little or no technical uncertainty. Each of the several machine concepts generated in Step 4 will have specific advantages and specific disadvantages.

If the best choice is not immediately obvious, then a systematic technical evaluation should be performed. The design constraints listed in Table 4-4 are

TABLE 4-6. Evaluation Matrix.

Criterion	Relative Weight	Alternative Concepts					
		A		B		C	
		score[1]	w.s.[2]	score	w.s.	score	w.s.
1. Meets production rate	20	10	200	10	200	5	100
2. Jamming tendency	20	5	100	8	160	10	200
3. Access to workstations	10	8	80	10	100	4	40
4. Accuracy	20	8	160	10	200	10	200
5. Off-the-shelf comp. vs. design from scratch	5	10	50	2	10	10	50
6. Safety of operator	15	10	150	10	150	10	150
7. Cost	10	10	100	2	20	6	60
			840		840		800

[1]Score 0 to 10. 0 = Totally unacceptable; 10 = meets all requirements without exception.
[2]Weighted score = score times relative weight.

also design criteria against which one can measure the degree of acceptability of the concept. This can be done using a matrix as shown in Table 4-6, first to weight the relative value of each individual criterion, and second, to score each design concept as to how well it meets each criterion. The resultant weighted scores for each design concept under consideration can then be used as a basis for selection of the best design concept.

4.3.6. The Machine Concept Design—Concept Selection *policies*

Keep in mind that the machine concepting stage usually takes place before a proposal is written, hence before the design-and-build job is firmly in hand. One unfortunate consequence of this fact is that there may be reasons other than purely technical ones for selecting a particular design concept.

The most commonly encountered nontechnical selection factor is machine cost. Most serious machine buyers have a budgeted amount to spend for a machine, and it is a waste of everyone's time to propose something far beyond that budget.

The second most common factor is that of required machine completion schedule. The delivery of any of the principal machine components is usually beyond the control of the builder, and thus becomes a major constraint on how quickly he can build and deliver the machine. It serves no purpose to propose a very elegant machine concept to a customer if it can only be delivered to him six months after his requirement.

There are numerous other nontechnical considerations in selecting the best machine concept to propose. Factors such as customer bias, financing arrangements, use of certain vendors, etc., fall into a category I call "the big picture." It is best that the designer have some feeling for the big picture as he concepts the machine—not to the extent of compromising his efforts to design the best possible machine, but to avoid what may be a political blunder which renders even the most elegant machine design irrelevant.

The final concept selected will generally be the best technical alternative with due consideration of all nontechnical factors taken into account. At this point a proposal can be written which describes the essentials of the machine, the cost to the customer for the machine design and build, and the schedule by which the machine will be delivered. The entire concepting process will rarely take more than ten working days to complete, with two or three days being more common. It is very expensive to dedicate a designer to concepting machines for proposals and it is not good business to do so unless there is a high probability of getting the job.

4.3.7. The Design Layout

When the machine design job is actually in hand, the first activity of the preliminary design stage (see Section 3.2.4) is to prepare a machine layout on the drafting board to full or reduced scale. Since most of the major machine components have been identified during the concepting phase, there is not much catalog searching to be done. It may be necessary to refer to vendor's data sheets or to contact the vendor directly to obtain full component dimensioning for the purposes of the layout.

Constructing the machine layout is like assembling a puzzle. Components which have been specifically identified have sizes and dimensions which must be physically attached to other components. If an off-the-shelf machine chassis is being used, then all work stations, parts feeders, etc., are simply added at the required locations. If a chassis must be constructed, then it is built up so as to support work stations and material transfer subsystems in proper relationship with one another.

If it can be said that there is an analysis stage in the design, then it occurs during preliminary design. Machine loading must be determined (see Section 6.1) to size and select the proper actuators. Critical structural members must be analyzed for stress loading, deflections under load, dynamic performance (i.e., vibration or resonances), and fatigue life. Control system inputs, outputs, and logic requirements must be derived and the control scheme outlined (see Section 9.1.4). During this phase, it is also wise to consider the effect of production rate requirements on machine efficiency and predicted performance.

A general rule to follow in developing the layout is to leave as much space as is practical under consideration of work transfer requirements, speed of operation, and overall machine size constraints for details which are yet to be designed. Often the detailing job is shared by several designers and/or draftsmen, particularly on a large machine, and it is not wise to complicate the detail design job by restricting space. Detailers take their "envelopes" directly from the layout drawing.

If there were any customer doubts about how or how well the machine would operate after the concept stage (although there should not have been) then these doubts or questions should all be fully answered after the layout phase. Progression of the design to completion should be a straightforward effort in the minds of everyone concerned.

4.3.8. The Design Details

This phase of the design process is perhaps the most mechanical of all phases. In essence the machine has been completely designed. However, it would be difficult indeed to build the machine from the layout drawing, even if it shows all views. The object, then, of the detail design phase is to generate working drawings from which all nonpurchased machine parts can be fabricated.

Traditionally, machinery details are all drawn of the same roll drawing. Government jobs usually require drawing packages to military specifications, which call for each part to be drawn on a separate sheet. Roll detailing is particularly efficient if the design house is also the build house, and all parts fabrication is to be done in house. Sheet detailing is necessary if the parts fabrication is going to be subcontracted out, possibly to several vendors.

TABLE 4-7.   Detail Design Tasks.

- Prepare individual detail drawings for all fabricated machine parts.
- Specify and document altered (purchased) items, if any,
- Specify and document interfacing and fastening of purchased machine components to the machine assembly.
- Perform tolerance study to assure proper machine references, alignments, and clearances.
- Select, design, and locate adjustment points required to fixture machine.
- Calculate stresses in machine members, holding brackets, pressure vessels, etc., and select material and size members.
- Specify material heat treatment for strength and hardness.
- Specify surface finishes, preparation, coating or plating for corrosion protection, wear resistance, etc.
- Write machine maintenance procedures.
- Prepare wiring diagrams, installation instructions.
- Write control programs (algorithms) if computer controlled.
- Document operation and troubleshooting of machine.

Detailing is a very time-consuming effort and requires a great deal of engineering talent. The fundamentals which are usually taught as "Design of Machine Elements" are put into practice here. Although each job has its own unique requirements, Table 4-7 lists many of the more common detail design tasks.

Control details must be completed during this phase also. If computer controls are to be used, then the program must be written from the block diagrams generated during the preliminary design phase. Electrical schematics and wiring diagrams must be generated so that electricians and technicians can wire up the machine. If the controls are being designed by someone other than the machine designer (as is the usual case) then the two must collaborate on practical details including mounting and access to control panels, physical location of wires, cable and conduit, ergonomics of readout locations and manual control devices, etc.

5. MACHINERY ECONOMICS

5.1. GENERAL CONSIDERATIONS

Although the design-and-build business is competitive, it is not cutthroat in the sense that price is the primary factor in buying a special production machine. The customer needs an automatic machine which will function to specification in a reliable manner for many years. A special automatic production machine almost by definition implies large (greater than 500K per year) production volume, and this invariably means large product dollar totals. In short, the customer expects to pay top dollar for a top-quality, reliable machine.

If price is no object, why not, you may ask, perform a cursory cost estimate, then double or triple that amount to quote the customer? There are two primary reasons:

First, the customer is not totally naive (after all, he is in the business of producing some product on a large scale). He generally knows the risks and costs associated with the proposed machine, and most likely will not stand for being cheated regardless of his need for the machine.

Second, there exists a machine price above which it is economically not feasible for the customer to automate. This maximum price depends on the customer cost savings anticipated from automation.

As described in the section concerning project concept design and proposal preparation, accurate estimation of the time and material required to design and build a special machine is of extreme importance. The overriding concern is not that one of "landing the job" among competition, but one of *not* losing money when the job is indeed won. The design and build of special machinery is an unforgiving business for those who underestimate project financial requirements or technical risks. Since the machine is one of a kind, there is little chance to recoup initial losses through sales of later copies of the machine. If the project cost overrun stems from a serious technical design error (i.e., the machine will not work to specification as designed) then the magnitude of the loss to the builder runs 70–100% of the total machine price. The design and build of a special machine is basically a one-shot deal. You design the machine, fabricate the parts, assemble the machine—and hope you can make it work to specification. There is no second chance—at least while the customer is picking up the tab.

It thus becomes clear why the machine designer must not only be technically accurate in concepting the machine—he must also be accurate in estimating the cost (manhours + material) to design, build, and debug the machine. The financial consequences to the builder are potentially enormous.

Before getting into the details of costing a design-and-build project, it is instructive to consider the various types of arrangements typically entered into between machine builder and machine purchaser.

At one end of the spectrum are *firm fixed price* contracts. Here, the customer agrees to pay a fixed amount to the machine builder upon delivery and successful demonstration that the machine meets performance requirements. Clearly the customer is taking no risk (except that he may never get the machine he wants) and the builder is assuming 100% of the technical and financial risk. The customer knows this, thus he expects and tolerates a higher fixed price tag on the machine than he would if he had agreed to share the risk somewhat. If all goes according to plan, the design is straightforward and the machine goes together with no glitches (and works), the builder stands to make a great deal of profit on the project. The flip side is, of course, that things rarely go so smoothly, and the builder stands the risk of being stuck with a very expensive piece of machinery (which he has paid for) which he can not sell because it does not meet specification. Or, as is more usually the case, the builder spends a great deal of money making the machine work, resulting in little or no profit, perhaps a loss on the project. No business, large or small, can tolerate very many such "losers" before shutting the doors.

On the other end of the spectrum is the *cost reimbursement* or *time and materials* (T&M) type of contract. Here the customer agrees to pay the builder for actual manhours spent on the project (at an agreed upon rate per hour) and for actual materials and purchased part costs (plus an agreed upon markup for handling). Here the builder cannot really lose money because he gets paid for everything he spends, and his profit is built into the agreed upon rate paid by the customer. It has been argued that the builder has no incentive to finish the machine on time and within budget under these circumstances. Clearly, all the builder stands to lose for poor performance is his reputation.

Because both of the above types of contracts in their extreme form contain serious negatives from the point of view of either the builder or the customer, hybrid machine build contracts, containing elements of each type are common:

Cost Plus Fixed Fee (CPFF). This type of contract is common in government work, and since the government does little production, they rarely buy automatic machines. Large government contractors do, however, let subcontracts to special machine design and builders quite often, and these subcontracts are subject to the same rules as the prime (production) contract. CPFF contracts reimburse the builder for actual costs incurred (at cost, not at an agreed upon rate) to complete the machine plus a negotiated fixed fee representing the builder's profit. As overruns occur, the basic costs are paid by the customer, but the builder profit margin (as a percent of the total job) drops. Since CPFF profit margins are low to start with (why not? the builder is taking absolutely no risk) there is no way to make a lot on the job.

Time and Materials, Not-To-Exceed. Here, work is performed on a T&M basis, but with an upper total price limit, beyond which the customer will not pay any more. From the builder point of view, this is a fixed price contract. From the customer point of view, this type of contract represents a chance to pay less than the fixed price ceiling if the machine costs less.

Fixed Price Design, T&M Build. The preliminary and detail design phases of a machine build contract can usually be very accurately estimated, since delivery consists of a drawing package which, if not fully approved by the customer, can be modified rapidly and at minimal cost. Therefore a fixed price deal makes sense. The customer has no control over the cost of the machine materials—they are determined by the design. Often the focus of much dispute in machine design and build is the debug phase—making the machine perform to specification. The customer naturally wants a machine which performs to 100% of specification—cycle time, reject levels, etc. But insistence on this point can run up the machine price very quickly. A T&M basis for debugging allows both parties to minimize financial risk—the builder gets paid for all his debugging efforts, and the customer has the option to stop accumulating costs by accepting something less than 100% machine performance, or by taking over the debugging himself. The latter is a viable option, since fine tuning of the machine can run months or years into the actual production on the machine.

Fixed Price Design and Build, T&M Debug. This is a slight modification of the preceding arrangement, wherein the machine is designed, built, and initially cycled in the customer's plant. In other words, the customer receives a working machine for a firm fixed price. Additional debugging of the machine takes place on a T&M basis. Since debugging a relatively complex machine can cost about 25% of the total price (one can never estimate exactly what the cost will be in advance), it is to both the customer's and the builder's advantage to work in this manner. This type of contract (or the previous one) probably results in the most equitable deal for both builder and customer.

Another important consideration, particularly for the small builder, is method of payment. Special machines are expensive items, and there is quite a bit of money tied up in the machine at almost every stage of its construction. The ideal situation for the builder would be to obtain 100% of the machine price up front. The ideal arrangement for the customer would be to withhold 100% of the price until the machine was fully operational in their plant. Since the financial risk of either of these extremes is unacceptable to the other party, a middle ground is usually negotiated.

Some acceptable arrangements follow:

1. *Periodic Cost Reimbursement.* Builder invoices customer for actual costs incurred (or at agreed upon billing rate) during some specified time period. Monthly invoices are common, biweekly invoicing is sometimes possible, particularly when builder has a tight cash flow situation. The customer is usually right on top of progress, since he is paying for time spent without a hard deliverable item.

2. *Progress Payments Against a Fixed Price Contract.* Builder invoices customer upon completion of certain key milestones as they are completed. Typical milestones include: preliminary design completion, detail design completion, purchase parts on order, machine assembled. Note that the customer does not have the completed machine in hand, but does have hard items (e.g., design package, machine components) against which payments are made. If the builder cannot deliver the complete machine, presumably the customer can use these milestone items to get the machine finished up elsewhere (albeit at somewhat greater cost than originally anticipated).

3. $\frac{1}{3}$ *Up Front,* $\frac{1}{3}$ *On Delivery,* $\frac{1}{3}$ *After Proof.* This is a common commercial arrangement for design and build of special machinery. Other percentages may be negotiated, as may other project milestones. The $\frac{1}{3}$ on delivery is usually payable once the machine is installed and operational in the customer's plant, with the final $\frac{1}{3}$ withheld until the machine is fully debugged and producing to specification.

Whatever the contract basis for building the machine may eventually be, and whatever the terms of payment may be, it is essential that the terms be clear to both parties.

5.2. PROJECT COSTING

Design and build work is usually costed commercially at a set hourly rate for labor and a materials cost plus markup. The hourly rate, which may vary for design time, engineering time, machining time, assembly time, etc., is set so as to cover the builder's direct costs, indirect costs (overhead), and profit. A typical and acceptable hourly rate is three times actual labor or salary. For example; if the average hourly rate in the machine shop is $10.00, an acceptable shop rate would be $30.00 per hour.

Costing a job on the basis of set rates for various categories of labor is all right as long as the rates are periodically checked by your accountant, and found to still be profitable to the company. However, the costing method described above is not acceptable to the government or to government contractors who subcontract their design and build. A more detailed costing procedure is required, one which gives the customer more insight into your company's financial operation than you would probably care to divulge; however, it comes with the job.

It is not a bad idea to cost all jobs in the manner to be described, even commercial machinery design and build. The customer need not know the details, he will only see the bottom line in your quote to him. The detailed costing exercise will, however, give you a great deal of insight into the job you are about to undertake, and most likely help you control it (i.e., keep costs from getting out of hand).

5.2.1. Costing Direct Labor

The first step in planning a design and build project is to break the job down into its most basic required tasks and subtasks. This can generally be done best by the machine concept designer because he knows what the problem areas will be, which tasks will require more time to complete than others, etc. He can thus apply actual manhour requirements to the individual tasks. Manhour requirements for shop time, assembly time, debug time, etc., if not known by the design engineer, can be obtained with the help of the shop estimator, assembly foreman, etc. The final result of this effort is a manpower loading chart. Fig. 5-1 depicts a hypothetical manpower loading chart.

Once manpower loading is defined, costing direct labor is relatively simple. Actual labor category salary rates are applied to the estimated number of hours to arrive at a direct labor cost for the project. Fig. 5-2 summarizes the direct labor costing of the assumed project manpower requirements.

One common alternative to breaking down the project and preparing a manpower loading chart is to make gross labor estimates in one of the following ways:

1. Estimate the job will take two engineers, a designer, two draftsmen, three machinists, and two technicians a total of five months to complete. Total the salaries of these people assigned to the probject for the required time period to get direct labor cost.

	Engineer 1	Engineer 2	Sr. Designer	Draftsman 1	Draftsman 2	Machinist 1	Machinist 2	Inspector	Assy. Tech. 1	Assy. Tech. 2	Mfg. Eng.
PHASE I. DESIGN											
Task 1. Preliminary Design											
a. Layout			80								
b. Power Train	80										
c. Control System		80									
d. Component Selection	40										
Task 2. Detail Design											
a. Drive Details	40		20	120							
b. Electrical Details		40	20		120						
c. Parts List	20	20	40	40	40						
d. Checking	80	80	80	40	40						
PHASE II. BUILD											
Task 1. Order and expedite	80	80	80								
Task 2. Machining						240	240	80			
Task 3. Assembly											
a. Mechanical	80								160		
b. Electrical		80								160	
c. In-house setup and cycle	20	20				40			40	40	
d. Break down and ship									20	20	
e. Assembly (customer's)									20	20	
PHASE III. DEBUG—CUSTOMER PLANT											
Task 1. Mech. Adjustments	80								80		80
Task 2. Elec. Adjustments		80								80	
Task 3. Run 1000 Parts	20	20	20						40	40	40
Task 4. Contingincies	80	80	80			40	40		80	80	80
TOTAL MANHOURS	620	580	420	200	200	320	280	80	440	440	200

Fig. 5-1. Manpower loading chart.

Mechanical engineer	620 Hrs @ $18.00	$11,160
Electrical engineer	580 Hrs @ $17.00	$ 9,860
Manufacturing engineer	200 Hrs @ $16.00	$ 3,200
Sr. designer	420 Hrs @ $15.00	$ 6,300
Mech. draftsman	200 Hrs @ $10.00	$ 2,000
Elect. Draftsman	200 Hrs @ $10.00	$ 2,000
Machinist 1	320 Hrs @ $13.00	$ 4,160
Machinist 2	280 Hrs @ $12.00	$ 3,360
Inspector	80 Hrs @ $11.00	$ 880
Assembly Tech. 1	440 Hrs @ $10.00	$ 4,400
Assembly Tech. 2	440 Hrs @ $ 8.00	$ 3,520
		$50,840

Fig. 5-2. Direct labor costing for hypothetical example.

2. Recall that a very similar machine was built last year for a direct labor total of $50,000. Add 10% for raises, etc. and quote direct labor on this job at $55,000.

Since every special machine is unique and has its unique problems, it is best to try to spend the time to cost out labor requirements precisely as possible. Cost overruns on jobs are rarely due to materials cost overruns or materials cost estimation error. Cost overruns are almost always due to underestimated labor requirements, and four times out of five, labor required to fully debug the machine.

5.2.2 Overhead on Direct Labor

Overhead includes all costs not directly attributable to the specific project, but nevertheless necessary for the job to get done. Since a design-and-build company is basically a service company, selling its time to customers, most indirect expenses (not all) grow in proportion to direct labor expenses. This being the case, the ratio of overhead expenses to direct labor expense remains fairly constant. If direct labor expense on a particular job is known, it is an easy matter to allocate the proper overhead to that job by simply multiplying direct labor cost by the overhead rate expressed as a fraction of direct labor.

The overhead rate is calculated from historical information by dividing the total of all overhead costs for a given period of time (called the overhead pool) by the actual direct labor costs for that same period of time. Thus:

$$1983 \text{ overhead rate} = \frac{\text{overhead pool } 1982}{\text{direct labor } 1982}$$

The overhead pool consists of the dollar sum of all overhead expenses for the chosen time period, and commonly includes the following:

- Engineering management
- Shop foremen
- Marketing labor
- Nonproductive time of direct people
- Fringe benefits
- Marketing budget (advertising, etc.)
- Lights, heat, power
- Facilities rent or amortization
- Drafting department supplies
- Shop supplies

- Facilities depreciation
- Product development
- QC department

Overhead labor usually makes up the largest part of overhead expense.

Typical overhead rates for a well managed design-and-build firm run between 75% and 150%. Higher rates are often seen, particularly in highly marketing oriented firms, R&D firms, and those firms which feel obliged to load up the nonproductive staffs with personnel.

Assuming an overhead rate of 75% on direct labor, the overhead cost of the example of Fig. 5-2 will be $38,130 (0.75 × $50,840).

5.2.3. Direct Materials Costs

Direct materials cost for the project under consideration requires a list of all purchased components (motors, controllers, drives, etc.) and all raw material required for fabricated machine parts. In addition, all machine components subcontracted out (anodizing, plating, heat treating, welding, etc.) are considered as direct materials expense. Rarely is a shop so well outfitted that it has facilities to perform all processes required in fabricating and building a machine.

Some companies apply a separate overhead to direct materials costs. This overhead rate is calculated exactly like the overhead rate on direct labor, but the materials overhead pool consists of indirect expenses related to material handling such as: purchasing department, expediting, material storage, transfer, control, etc. A typical rate would be 5–20%.

For the purposes of our example, we will assume the material overhead expenses are included in the overhead rate on direct labor. While the inherent assumption that indirect costs associated with materials are somehow proportional to direct labor is not always correct, most companies avoid the excess bookkeeping and lump all overhead costs into one rate.

Assume that materials costs have been estimated to be $80,000.

5.2.4. Other Direct Costs

If specific costs can be associated to the job, they should be charged as direct costs. Typical direct cost include:

Travel: 4 man-trips to customer's plant	$ 3600
Rigging: Deliver machine to customer's plant:	$ 8800
Special Test: Rental of environmental test equipment and facilities:	$ 4900
TOTAL OTHER DIRECT COSTS	$17300

5.2.5. General and Administrative (G&A)

Indirect expenses which are not necessarily proportional to direct labor, but are nevertheless necessary company expenses are covered by a factor known as G&A. The G&A pool includes: officer's salaries, secretaries' salaries, accounting salaries and expenses, computer, insurance (product liability insurance for machine builders can be quite expensive), legal fees, etc. In general, these costs will not increase in proportion to increases in the productive work force (direct labor).

The G&A rate is applied to the total of direct labor (DL), overhead on direct labor (OH), direct materials (DM) and other direct costs (ODC). It is calculated from historical data in exactly the same manner as is overhead:

$$1983 \text{ G\&A rate} = \frac{\text{G\&A pool 1982}}{(\text{DL} + \text{OH} + \text{DM} + \text{ODC}) \ 1982}$$

Reasonable G&A rates range from 15% to 25%. For our example, the sum of DL + OH + DM + ODC is $186,270. The G&A cost at an assumed rate of 15% will be $27,940.

5.2.6. Profit

The name of the game is to make a profit on every machine design-and-build project the compay undertakes, Profit is simply the difference between the machine price (which the customer pays) and the machine cost which we have tried to estimate in the preceding sections.

The sum of DL + OH + DM + ODC + G&A represents the real total cost of the machine. All costs, including indirect costs, have been covered

DIRECT LABOR COST		
(see Fig. 5-2)		$ 50,840
OVERHEAD ON DIRECT LABOR @ 75%		$ 38,130
DIRECT MATERIALS COST		$ 80,000
OTHER DIRECT COSTS		
Travel	$3,600	
Rigging	$8,800	
Special Test	$4,900	
		$ 17,300
SUBTOTAL		$186,270
G & A @ 15% of subtotal		$ 27,940
TOTAL MACHINE COST		$214,210
PROFIT @ 10%		$ 21,421
TOTAL MACHINE PRICE TO CUSTOMER		$235,631

Fig. 5-3. Cost and pricing summary for hypothetical example.

accurately if DL, DM, and ODC have been estimated correctly, and if OH and G&A rates are indeed accurate.

To this total machine cost must be applied some amount for profit. Government contracts and subcontracts often limit the amount of profit the builder can make. Cost Plus Fixed Fee (CPFF) contracts limit profit to 8% above total cost. The 8% is based on the estimated cost, so if the actual machine cost overruns the estimate, the fixed fee (profit) remains constant. Firm Fixed Price (FFP) contracts limit profit to 15% maximum, but typically allowed profit is only 10%. Here, of course, the builder eats any cost overruns, hence the additional allowed profit margin for the extra risk. On the positive side, the builder can increase his real profit margin by completing the machine at less than the estimated cost.

The allowable profit on commercial machine build contracts is not limited by law, so higher profit margins are always applied. Typical reasonable before-tax profit margins run 20–30%. It should be apparent why many firms refuse to perform government contract work, concentrating solely on commercial markets.

Figure 5-3 summarizes the costing exercise followed so far, and prices the machine with a 10% profit margin, assuming a Government Fixed Price contract.

5.2.7. Alternative Costing Technique

Many shops quote Time and Materials jobs as well as Fixed Price jobs based on a shop rate for the various services required. A typical shop rate is three

Category	Actual Rate	Average Rate	(Actual × 3) Shop Rate
Mechanical engineer	$18.00		
Electrical engineer	$17.00	$17.00	$51.00
Manufacturing engineer	$16.00		
Sr. Designer	$15.00		
Mech. draftsman	$10.00	$11.67	$35.00
Elect. draftsman	$10.00		
Machinist 1	$13.00		
Machinist 2	$12.00	$12.00	$36.00
Inspector	$11.00		
Assembly tech. 1	$10.00	$ 9.00	$27.00
Assembly tech. 2	$ 8.00		

NOTE: This example assumes no other personnel in the various departments. If there were others in the Engineering, Drafting, Shop, and Assembly Departments, then the average rate would be the average of everyone in the department, and the shop rate would be three times that amount (if three was an acceptable factor).

Fig. 5-4. Calculation of shop rate for direct labor categories.

accountant

LABOR
  Engineering: 1400 Hrs @ $51.00 $71,400
  Design: 820 Hrs @ $35.00 $28,700
  Shop: 680 Hrs @ $36.00 $24,480
  Assembly: 880 Hrs @ $27.00 $23,760
                                           ————
                                                      $148,340

MATERIALS
  Machine purchase parts: $80,000
  Other direct costs: $17,300
                                           ————
                                           $97,300
  Plus: 15% Markup on materials and expenses $14,595
                                           ————
                                                      $111,895
                                                      ————
  TOTAL MACHINE PRICE $260,235

NOTE: Machine *cost* to build is still $214,210 as calculated in Fig. 5-3. Therefore this method of costing results in a built-in profit of $46,025. or 21.5%.

Fig. 5-5. Alternate method of pricing the hypothetical machine.

times direct labor cost. A particular job would then be estimated by multiplying the required number of hours per category times that category's shop rate to arrive at a total price. Direct materials costs are summed up as before, and are marked up by 15–20% to cover handling.

Figure 5-4 shows the calculation of individual labor category rates based on three times actual salary. Fig. 5-5 summarizes the pricing of our hypothetical job using these rates and a 15% markup on direct materials and direct expenses. Compare the total machine price to that arrived at in Fig. 5-3. The difference is all profit, since actual machine costs total $214,210, as shown in Fig. 5-3.

6. ACTUATOR AND DRIVE SYSTEM PRINCIPLES

6.1. MACHINE FORCES, TORQUES, AND POWER

> ma·chine':…2. Technically, a combination of mechanical parts, such as levers, gears, pulleys, etc., serving to transmit and modify force and motion in such a way as to do some desired work.[1]

Every special production machine performs work on an object, known as the *product* or the *workpiece*. The magnitude of work performed varies considerably depending on the machine operation in question but in all cases is manifested by the exertion of a force F through a distance x or by the exertion of a torque T through an angle θ.

It is a very important conceptual design aid to understand that all machine work performed on a workpiece is implemented by linear (straight-line) or rotational motion or a combination of both at the machine/work interface. In addition, all of these machine output motions and the forces and/or torques associated with them are initiated by prime movers which themselves produce either linear or rotary motions.

It should serve as some consolation to the machine designer, then, that of all the thousands of possible machine requirements, all operations can be reduced to applying torques through some angle and forces through some linear distance, using actuators with outputs which are either linear motion or rotary motion.

Machine power output is defined as the rate at which work is performed by the machine/work interface. The machine power output requirement is often called the *machine load*. The first step in concepting a machine is to concisely identify all of the machine requirements.

Similarly, as the designer gets down to the nuts and bolts of designing the machine drive system, the first step is to calculate the expected machine load. By working backwards from the machine load at the machine/workpiece

---

[1] *Webster's Students Dictionary,* American Book Company, New York, 1959.

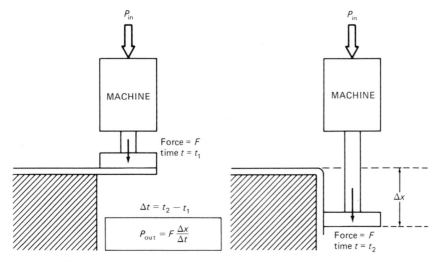

Fig. 6-1. Machine exerting linear output power to perform useful work.

interface, the designer must ultimately calculate the actuator load so that the appropriate drive actuator may be selected.

Fig. 6-1 depicts a machine with a power output P_{out} which is produced by exerting a constant force F through a linear distance Δx in a time Δt. This action is similar to that required by a hydraulic ram to bend a section of sheet metal. The power output of the machine can be written as

$$P_{out} = \frac{dW}{dt} = \frac{d(F \cdot x)}{dt} = F\frac{dx}{dt} = F\frac{\Delta x}{\Delta t} = F \cdot V, \qquad (6\text{-}1)$$

where $W =$ Work performed on workpiece defined as force F times distance
x (ft-lb)
$V = \Delta x/\Delta t =$ Velocity of ram (ft/sec).

Fig. 6-2 depicts a machine with power output P_{out} produced by exerting a constant torque T through an angle $\Delta\theta$ in a time Δt. This is the power which would have to be exerted by a machine tool to make a cut with a milling cutter. Usually we consider the machine output to be a torque on the tool shaft rather than a cutting force at the cutter/workpiece interface because the cutter is most likely driven by a motor which is delivering some output torque, although there is equivalence between the shaft torque and the cutting force acting at a moment arm equal to the cutting tool radius. The power output of

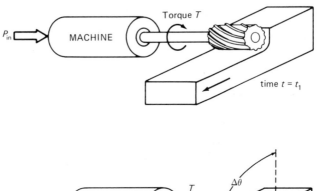

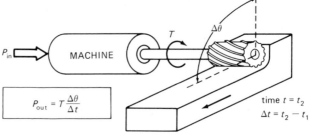

Fig. 6-2. Machine exerting rotary output power to perform useful work.

the machine, P_{out}, in ft-lb/sec, can be written as

$$P_{out} = \frac{dW}{dt} = \frac{d(T \cdot \theta)}{dt} = T\frac{d\theta}{dt} = T\frac{\Delta\theta}{\Delta t} = T \cdot \omega \qquad (6\text{-}2)$$

where W = Work performed in applying a torque T through an angle θ (ft/lb-radians)

$\Delta\theta$ = Angular displacement about output shaft (radians)

ω = Angular velocity about output shaft (radians/sec).

Fluid power is often implemented in machinery design by hydraulic or pneumatic cylinders and motors. The power required to move a volume flow rate Q through a load which results in a pressure drop Δp is (see Fig. 6-3):

$$P_{fluid} = Q \cdot \Delta p \qquad (\text{ft-lb/sec}) \qquad (6\text{-}3)$$

where Q = Fluid flow rate, ft³/sec

Δp = Pressure differential between fluid inlet and fluid outlet, lb/ft².

The common unit of mechanical power used to describe machine element capability is horsepower, and the common electrical power unit is watts (or kilowatts). Similarly, angular velocity is usually specified in revolutions per minute (rpm), pressure in lb/square inch (psi), and fluid flow in gallons per

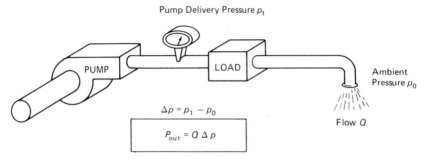

Fig. 6-3. Power associated with pumping fluid across a load.

minute (gpm). Table 6-1 presents the power equations (6-1) through (6-3) in several different ways, using more commonly encountered units.

6.2. THE CONCEPT OF A DRIVE SYSTEM

Fig. 2-1 depicted a generalized machine broken down into functional subsystems, i.e., work stations, material transfer, and control. This conceptual model allows the designer to think in terms of *what* the machine must do. Fig. 6-4 depicts a *power-flow* model for a generalized machine which allows the designer to visualize *how* the machine will perform the necessary function

TABLE 6-1 Equations for Calculating Power.

Linear Power, Eq. (6-1)	$P = 1.82 \times 10^{-3} \, F \cdot V$ (horsepower) $\frac{lb \cdot ft}{sec}$
	$P = 1.356 \, F \cdot V$ (watts)
where:	$F =$ Linear force in lb
	$V =$ Linear velocity in ft/sec
Rotary Power, Eq. (6-2)	$P = 1.9 \times 10^{-4} \, T \cdot n$ (horsepower) $ft \cdot lbs \cdot Rpm$
	$P = 0.14 \, T \cdot n$ (watts)
where:	$T =$ Torque in ft-lbs
	$n =$ Rotation speed in rpm
	$P = 1.52 \times 10^{-4} \, T \cdot \omega$ (horsepower) $T = I \alpha$
	$P = 0.113 \, T \cdot \omega$ (watts) $\frac{\omega}{t}$
	$T =$ Torque in inch-lbs.
	$\omega =$ Rotation speed in radians/sec.
Fluid Power, Eq. (6-3)	$P = 5.84 \times 10^{-4} \, Q\Delta p$ (horsepower) $\frac{gl}{min} \cdot psi$
	$P = 0.435 \, Q\Delta p$ (watts)
where:	$Q =$ Fluid flow rate in gallons/minute
	$\Delta p =$ Pressure drop in lb/in.^{2} (psi)

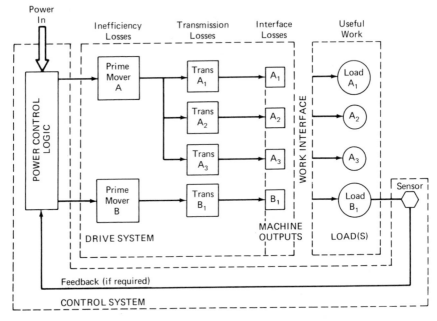

Fig. 6-4. Generalized power-flow model of a machine.

of applying power (i.e., do useful work) to the workpiece. The power-flow model of a machine has four basic and arbitrary subdivisions, outlined below.

1. *The drive system* receives as its input controlled power which it first transforms into that form required at the machine output or work interface. The two most common transformations are electrical to mechanical (motors or solenoids) and fluid pressure to mechanical (pneumatic or hydraulic cylinders). The transformation devices are referred to as *actuators* or prime movers.

The second half of the drive system is the *transmission,* which transmits power in its appropriate form to the machine output at the work interface.

There may be only one prime mover from which all required power is derived (by power take-offs). For example, a single motor might operate the work transfer subsystem as well as actuate the work station(s) by means of gearing, clutching, cams, etc.

There may also be more than one actuator, each with its own transmission, all independently controlled to provide useful work at the machine output.

2. *The work interface* or *machine output* is the machine element interfacing the workpiece at each station, as the work is transferred through the

machine, or as parts are moved in crossline transfer. It is the element which physically applies the output power to the work.

In the case of work stations, the work interface is called a *tool.*

In the case of work transfer, the work interface is called a *work carrier,* a nest, a pallet, a fixture, a gripper, or simply an actuator.

3. *The machine load* is the reaction power between the work and the work interface and is equal to the power transmitted to the work by the machine plus any losses that occur at the interface.

For most work stations, the machine load is a force (torque) applied through a distance (angle) at some speed. If heavy tooling is used, then there will be inertial components to the load.

For most material transfer operations, the machine load consists of inertial loads to accelerate the work, friction loads to move the work along, or gravity against which the material must be lifted.

4. *The control system* takes in uncontrolled power (usually line electricity, air pressure, or hydraulic pressure and flow) and delivers controlled power to each actuator (prime mover).

It also accepts control inputs from both within and/or outside of the machine. Internally generated control signals require *sensors* and by definition classify the control system as *closed loop.* Control signals from the outside include manual inputs and inputs from a larger system of which the machine may only be a part (e.g., in an automated factory).

The control system delivers controlled power to each actuator according to the timing and sequence dictated by the *control logic* or *control algorithm.* The control system decides what the machine should do and directs it accordingly based on each control input or combination of inputs.

Design of a machine drive system consists of sizing and selecting both actuators and transmission elements. As Table 6-2 indicates, there are very few basic actuators used in machinery design. It is interesting to note that all machine output motions are ultimately caused by actuators which provide either straight-line (linear) or circular (rotary) motion.

Fig. 6-5 shows the steps in power train design. As the figure indicates, the final objective of the design is to select actuators which not only provide sufficient power, but which supply this power (i.e., torque at the proper rpm; force at the proper velocity; proper flow across a pressure drop) with maximum efficiency. In order to select the right actuator, an analysis of the machine loads and drive train must be made so as to determine actuator load, or specific torque/speed (force/speed) requirements *at the actuator shaft.*

Section 6.3 presents methods for calculating basic machine loads, power

TABLE 6-2   Basic Actuators Used in Machinery Design.

| | Electric | Fluid Power | |
		Hydraulic	Pneumatic
1. Linear, intermittent	solenoid, motor screw	cylinder	cylinder
2. Rotary, Continuous	AC motor, DC motor	hydraulic motor	air motor
3. Rotary, intermittent	solenoid	hydraulic servo	cylinder mechanism (rotary indexer)
	step motor	cylinder mechanism (rotary indexer)	
	servomotor, motor clutch		

requirements to accelerate a machine or workpiece and frictional losses at the workpiece or in the transmission system. Section 6.3.5 summarizes the procedure for calculating actuator loads from machine loads. Section 6.4–6.6 discuss the specifics of the basic types of actuators, electric motor, fluid cylinders, and electric solenoids.

6.3.   QUANTIFYING MACHINE LOADS AND LOSSES

The first step in designing the main drive system or auxiliary drive systems in a machine is to determine the machine load. Step 2 of the concept design process (see Section 4.3.2) resulted in a complete listing of all of the required machine operations for both work stations and material transfer. It is necessary to calculate the work requirement for each of these operations and to quantify the machine output required as follows:

1. Torque (or force) required at machine output to perform specific task (use maximum expected forces);
2. Displacement, or stroke through which forces are applied;
3. Speed at which work must be performed; in the case of material transfer, time limitations to make the transfer;
4. Losses anticipated at the machine/work interface (friction, windage, heating, etc.);
5. Output power requirement based on 1–4 above.

Having identified force and stroke requirements for each operation, the designer must then design or select components for the power transmission system so as to provide the required output motions from a general type of

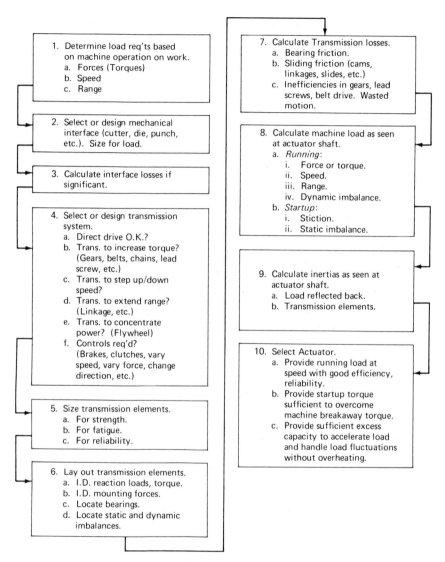

1. Determine load req'ts based on machine operation on work.
   a. Forces (Torques)
   b. Speed
   c. Range

2. Select or design mechanical interface (cutter, die, punch, etc.). Size for load.

3. Calculate interface losses if significant.

4. Select or design transmission system.
   a. Direct drive O.K.?
   b. Trans. to increase torque? (Gears, belts, chains, lead screw, etc.)
   c. Trans. to step up/down speed?
   d. Trans. to extend range? (Linkage, etc.)
   e. Trans. to concentrate power? (Flywheel)
   f. Controls req'd? (Brakes, clutches, vary speed, vary force, change direction, etc.)

5. Size transmission elements.
   a. For strength.
   b. For fatigue.
   c. For reliability.

6. Lay out transmission elements.
   a. I.D. reaction loads, torque.
   b. I.D. mounting forces.
   c. Locate bearings.
   d. Locate static and dynamic imbalances.

7. Calculate Transmission losses.
   a. Bearing friction.
   b. Sliding friction (cams, linkages, slides, etc.)
   c. Inefficiencies in gears, lead screws, belt drive. Wasted motion.

8. Calculate machine load as seen at actuator shaft.
   a. *Running*:
      i. Force or torque.
      ii. Speed.
      iii. Range.
      iv. Dynamic imbalance.
   b. *Startup*:
      i. Stiction.
      ii. Static imbalance.

9. Calculate inertias as seen at actuator shaft.
   a. Load reflected back.
   b. Transmission elements.

10. Select Actuator.
    a. Provide running load at speed with good efficiency, reliability.
    b. Provide startup torque sufficient to overcome machine breakaway torque.
    c. Provide sufficient excess capacity to accelerate load and handle load fluctuations without overheating.

Fig. 6-5. Drive train design procedure.

actuator (e.g., electric motor running at 1750 rpm). Once the power transmission is defined, efficiency losses due to friction must be calculated using the methods of this section. Finally, the actuator must be sized by considering machine load (as seen by the actuator), losses through the transmission system, and inertial loads which must be overcome to accelerate both the work and the machine components in the power train.

6.3.1. Overcoming Frictional Forces

Frictional forces must be overcome in machinery in both the work station and the material transfer subsystem.

Typical examples of friction in work stations include:

- Friction due to ram guides in a punching operation;
- Friction due to die locator pins in a stamping or forming operation;
- Bearing, slide, gear, lead screw, plunger friction in the power train between actuator and workpiece;
- Friction torque experienced in driving a screw in an assembly operation.

In the material transfer subsystem, frictional forces appear often as:

- Friction between workpiece and machine as work is moved by sliding;
- Friction between parts and parts feeding slides as they are fed to assembly position;
- Friction between belts or conveyors and work as work is transferred, often accelerating and decelerating rather quickly;
- Component (gears, slides, lead screws, bearing, etc.) friction in the material transfer drive train.

The basic machine actuators (or prime movers, as they are often called) must overcome friction while running, and in the case of the very common intermittent machine, they must be capable of overcoming static friction or "stiction" with every machine cycle. Friction may be of the Coulomb type, independent of machine member speeds, or of the viscous type, linearly dependent on machine member speed. Obviously the friction may be rotary or linear, depending on the nature of relative motion of the machine member.

Linear Coulomb Friction. In the static case, a mass of weight W will start to slide from rest when an applied force F:

$$\text{static} \qquad F = \mu_s\, W \cos\theta + W \sin\theta \qquad (6\text{-}4)$$

where: F = Force required to start the weight in motion (lb.)
W = Weight of the mass (lb)
μ_s = Static coefficient of friction (dimensionless)
$W \cos\theta$ = Normal force between the mass and an inclined surface at angle θ (lb)
$W \sin\theta$ = Component of the object weight parallel to the inclined surface which must be "lifted," this is not a frictional force (lbs).

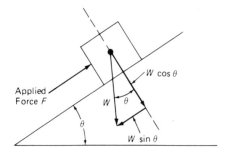

Fig. 6-6. Linear static friction.

Fig. 6-6 defines the case under consideration. Note that the frictional force between the inclined surface and the object is independent of the contact area between them.

The static friction calculation is made in machine design to:

1. Determine machine start-up forces due to static friction;
2. Determine frictional forces available from a belt or conveyor to move an object of a given weight through a given incline;
3. Determine the tendency for an object to slide or tip when started or stopped on a conveyor belt;
4. Determine the proper angle for a parts feed slide to assure parts do not "hang up" by friction.

Often the static coefficient of friction between two materials of a certain surface finish cannot be determined by reference to handbook values. A simple, quick experiment to determine the value consists of placing the actual part under consideration onto the slide (or a piece of the same material as the slide is to be made from) or belt to be used, and the inclining the slide until the part just begins to slide. The static coefficient of friction μ_s will be:

$$\mu_s = \tan \theta \qquad (6\text{-}5)$$

where θ = the angle at which the part just starts to slide.

The dynamic, or sliding coefficient of friction is generally 20–25% lower than the static coefficient of friction. Therefore, once the static coefficient of friction is overcome and a body starts to slide, the force required to keep that body sliding drops to 75–80% of its peak value. The force required to keep a body sliding at a constant velocity would thus be:

$$F = \mu_d W \cos \theta + W \sin \theta \qquad (6\text{-}6)$$

Where μ_d = Dynamic coefficient of friction (dimensionless).

TABLE 6-3 Coefficients of Friction for Common Material Combination ∽⁻26 - 25%

	Static	Dynamic
Hard steel on hard steel, dry	0.78	0.42
Hard steel on hard steel, greasy	0.11–0.23	0.03–0.12
Mild steel on mild steel, dry	0.74	0.57
Hard steel on graphite, dry	0.21	—
Aluminum on mild steel, dry	0.61	0.47
Teflon on steel	0.04	0.04
Copper on mild steel	0.53	0.36
Nickel on nickel	1.1	0.53
Brass on mild steel	0.51	0.44
Copper on cast iron	1.05	0.29
Aluminum on Aluminum	1.05	1.4

SOURCE: *Standard Handbook for Mechanical Engineers*, Baumeister & Marks, Seventh Edition, McGraw Hill.

Table 6-3 lists values for coefficients of friction for several combinations of materials. Example 6-1 demonstrates the practical application of static friction in a machine problem.

EXAMPLE 6-1

A conveyor belt moves molds from station to station, where they are filled with a castable epoxy, have reinforcement mesh inserted, and are force cured in an oven. Molds weigh 50 lb each and an additional 60 lb of material is eventually put into each mold. At maximum speed, the molds move at 80 ft/min. If a brake is applied and the conveyor comes to a complete stop from maximum speed in 0.2 seconds, what must be the static coefficient of friction between the belt and the molds to keep the molds from sliding?

SOLUTION: The force acting on the mold tending to keep it in motion during belt deceleration is:

$F = m \cdot a$.

$\left[N = kg \, m/s^2 \right]$

$$F_{inertial} = \frac{W}{g} \frac{\Delta V}{\Delta t} \text{ [Eq. (6-13)]}$$

$$= \frac{(110)(80)}{(32.2)(60)(0.2)}$$

$$= 22.7 \text{ lb.}$$

The friction force restraining the molds from sliding is:

$$F_{friction} = \mu W \ [Eq. \ (6\text{-}4)] = (110)(\mu).$$

If the molds are not to slide:

$$F_{friction} > F_{inertial}$$

$$110\mu > 22.7 \ lb$$

So:

$$\boxed{\mu > 0.21.}$$

Rotary Coulomb Friction. Static or dynamic coulomb friction associated with rotating shafts or brakes within a machine give rise to friction torques about a shaft axis. Typical sources of friction torque include bearings, bushings, and lead screws. In starting up a machine, static friction torques associated with all rotating shafts must be overcome. These static torques drop off to a steady dynamic friction torque level and must be continuously overcome by the driving actuator. In the case of an intermittent machine where shafting comes to rest then starts rotation repeatedly, the higher static friction torques must be continuously overcome.

Frictional torques in bearings or lead screws are often specified in the manufacturer's literature as torques (in inch-ounces), which is quite convenient. However, frictional values are also specified in terms of a coefficient of friction, from which the frictional torque must be calculated based on shaft diameter and shaft loading. The appropriate relationship is:

$$T_f = \mu D F_s \qquad (6\text{-}7)$$

where: T_f = Frictional torque (in.-lb)
$\quad \mu$ = Coefficient of friction for the bearing or screw
$\quad D$ = Shaft diameter at bearing bore (in.)
$\quad F_s$ = Side load (lb) on the shaft at the bearing or side load on a lead screw at the drive nut.

Table 6-4 lists some coefficients of friction for various bearing types.

Frictional torques within a machine can usually be minimized by good design practices, which generally means using rolling element (low friction) bearings on all drive shafting, and assuring proper lubrication for those bearings. To insure long life (essential in an automatic machine), the bearings

TABLE 6-4   Rolling Bearing Coefficients of Friction (Values with ideal lubrication).

Type of Bearing	Typical Coefficient of Friction (referred to bearing bore)
Ball Radial	0.0015
Ball Thrust	0.0015
Cylindrical Roller	0.002
Cylindrical Roller Thrust	0.008
Spherical Roller	0.002
Tapered Roller	0.002
Tapered Roller Thrust	0.002
Needle Roller	0.004

should be selected to be lightly loaded even under maximum machine loading. In a complex machine, the contribution to machine start-up torque due to static friction can be high. However, selection and sizing of the proper drive motor such that frictional torques are overcome is a straightforward procedure.

The real problem which coulomb friction can cause in a machine is heat build-up due to energy dissipation as heat. Coulomb friction is a nonconservative force, hence it continuously generates heat, as follows:

$$Q = 6.733 \times 10^{-4} \, T_f \, Nt \tag{6-8}$$

where Q = Heat generated in time t by a shaft rotating at N rpm from a source of friction torque T_f (BTU)
  T_f = Friction torque from a particular site, say a bearing (in.-lb)
  N = Rotational speed of the source of frictional torque (rev/min)
  t = Time that shaft runs at N rpm (hr).

Assuming that all the heat generated is absorbed by the steel in the vicinity of the friction source, that region will heat up by an amount ΔT:

$$Q = WC_p\Delta T \quad \approx \quad mc \, \Delta T \tag{6-9}$$

where: Q = Heat generated from friction source (BTU)
  W = Weight of material (steel) in contact with heat source (lb)
  C_p = Specific heat of material in contact with heat source
   = 0.11 BTU/lb-°F for steel
  ΔT = Temperature rise (°F) of the material in contact with the heat source.

Uneven heating in a machine, particularly a precision machine, can result in shaft misalignments due to thermal expansion. Since temperature build-ups take place gradually over time, and depend on environmental factors such as ambient air temperature in the vicinity of the machine and air currents around the machine, the misalignments and their effect on machine output can be quite erratic. Situations like this are very difficult to identify, isolate, and eventually fix.

Linear Viscous Friction.   Coulomb friction is characterized by the fact that the frictional force dissipated by a body in motion is independent of the speed with which one body slides with repect to another. Additionally, the frictional force is dependent on the normal force pressing the bodies together and independent of the contact area. Viscous friction is directly proportional to the relative sliding speed of one body with respect to another, the area of surface contact, and the viscosity of the fluid between the two bodies.

Fig. 6-7 depicts the case where a body is sliding with viscous friction over a flat surface. The force required to keep the body sliding at a constant velocity is given by:

$$F_v = A\mu V / \delta \qquad\qquad (6\text{-}10)$$

where: F_v = Applied force keeping block sliding (lb)
μ = Fluid viscosity (lb-sec/ft^2)
A = Contact area (ft^2)
δ = Clearance gap between the block and the surface which is filled with fluid (ft)
V = Velocity of the block over the surface (ft/sec).

Note that the viscous force does not depend on the weight of the block except as that weight might effect the clearance gap δ.

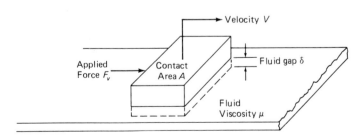

Fig. 6-7. Linear viscous friction.

Rotary Viscous Friction.   The linear viscous friction case discussed above takes place in a hydrostatic slide, usually used to support and slide very heavy weights. The more common rotary case occurs when a shaft rotates within an enclosure filled with oil. For the case of a cylinder rotating in an oil filled enclosure, the frictional torque at the cylinder outer diameter is:

$$T_v = 0.0823 \, \frac{\mu l D^3}{\delta} \, n \tag{6-11}$$

where: T_v = Viscous friction torque about the shaft axis (ft-lb)
μ = Viscosity of the fluid in the clearance gap (lb-sec/ft^2)
l = Length of the gapped area (ft)
D = Diameter of the shaft at the clearance gap (ft)
δ = Gap dimension (radial clearance) (ft)
n = Shaft rotational speed (rev/min).

Fig. 6-8 depicts the case described by Eq. (6-11). Fig. 6-9 shows the case of viscous friction on a circular face rotating with fluid in the axial clearance gap. The viscous friction torque acting on the shaft is

$$T_v = 0.0103 \, \frac{\mu D^4}{\delta} \, n \tag{6-12}$$

where variables and units are identical to those of Eq. (6-11).

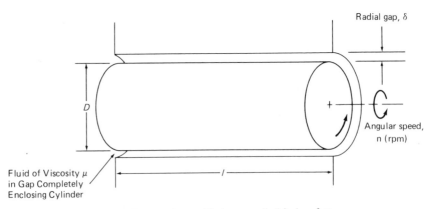

Fig. 6-8. Rotary viscous friction on cylindrical surface.

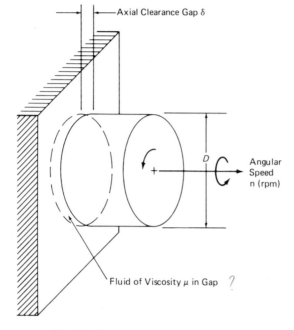

Fig. 6-9. Rotary viscous friction on disk.

6.3.2. Inertial Forces in Machinery

Machine elements of finite mass tend to resist motion when at rest, and exhibit a tendency to continue moving when in motion (Newton's First Law). It therefore requires more force, torque, or power to accelerate a machine member from rest than it does to maintain the machine in operation. In an intermittent machine, the high speed starting and stopping of various machine components may be the determining factor in selecting and sizing actuators.

The relationships for inertial forces and torques which follow are based on the general relations:

$$F = \frac{W}{g}\frac{d^2x}{dt^2} = \frac{W}{g}\frac{dV}{dt} \qquad ma \tag{6-13}$$

$$T = I_x\frac{d^2\theta}{dt^2} = I_x\frac{d\omega}{dt} \tag{6-14}$$

where: F = Force required to accelerate a body of weight W at a linear rate dV/dt (lb)

W = Weight of the body (lb)

g = Gravitational constant converting body weight to body mass
$= 32.2$ ft/sec^2

T = Torque required to accelerate a body with inertia I_x about the x axis at a rate $d\omega/dt$ (ft-lb)

I_x = Moment of inertia of body about the rotational (x) axis (ft-lb-sec^2).

ω = Angular velocity of the body (rad/sec)

The constant linear force required to accelerate a body of weight W from rest to a speed V (ft/sec) in time t (sec) is:

$$F = 0.031 \ WV/t. \tag{6-15}$$

The constant linear force F required to move a body of weight W from rest through a distance x (ft) in time t (sec) is:

$$F = 0.062 \ Wx/t^2. \tag{6-16}$$

The constant torque T (ft-lb) required to accelerate a body with inertia I_x (ft-lb-sec^2) from rest to an angular velocity N (rev/min) in time t (sec) is:

$$T = 0.105 \ \frac{I_x}{t} \ N \qquad\qquad \tag{6-17}$$

The constant torque T required to turn a shaft with inertia I_x (ft-lb-sec^2) from rest through an angle θ (degrees) in time t (sec) is:

$$T = 0.035 \ \frac{I_x \theta}{t^2}. \tag{6-18}$$

Fig. 6-10 shows moments of inertia (I_x) for various machine member shapes commonly encountered in machinery design. To calculate the motor torque required to accelerate a shaft and flywheel from rest to N rpm, one would use Eq. (6-17) with moments of inertia calculated from Fig. 6-10(a) (see Example 6-2). However, as is often the case, the flywheel may not be driven directly, but through a gear train of some type. In this case, the motor does not "see" the flywheel moment of inertia as calculated in Fig. 6-10; rather, it sees a different flywheel moment of inertia said to be "reflected"

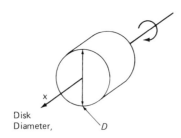

$$I_x = 0.00388\ WD^2$$
$$(\text{lb–ft-sec}^2)$$

W = Weight (lbs)
D = Diameter (ft)

(a) Disk.

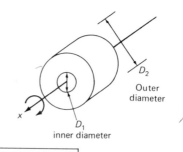

$$I_x = 0.00388\ W\left(D_1^2 + D_2^2\right)$$
$$(\text{lb–ft-sec}^2)$$

W = Weight (lbs)
D_1 = Inner diameter (ft)
D_2 = Outer diameter (ft)

(b) Thick walled cylinder.

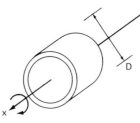

$$I_x = 0.00776\ WD^2$$
$$(\text{lb–ft-sec}^2)$$

W = Weight (lb)
D= Diameter (ft)

(c) Thin walled cylinder.

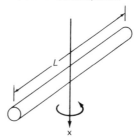

$$I_x = 0.00259\ WL^2$$
$$(\text{lb–ft-sec}^2)$$

L = Length (ft)
W = Weight (lb)

(d) Thin rod.

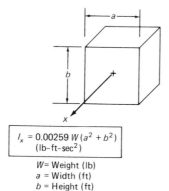

$$I_x = 0.00259\ W\left(a^2 + b^2\right)$$
$$(\text{lb–ft-sec}^2)$$

W= Weight (lb)
a = Width (ft)
b = Height (ft)

(e) Box.

Fig. 6-10. Moments of intertia for common bodies.

back through the gear train. For a gear train with gear ratio G and gear diameters D,

$$G = \frac{D_{\text{driven}}}{D_{\text{driver}}} \qquad (6\text{-}19)$$

the reflected moment of inertia $I_{\text{reflected}}$ is expressed as

$$I_{\text{reflected}} = \frac{I_x}{G^2} \qquad (6\text{-}20)$$

where I_x is the moment of inertia of the load. Example 6-3 considers the problem of Example 6-2 with a simple gear train inserted between motor and flywheel.

EXAMPLE 6-2

A DC electric motor is controlled to deliver a constant output torque of 240 inch pounds from rest up to a maximum motor speed of 500 rpm. (A motor which operates in this mode is called a *DC torque motor* or *DC torquer*). The polar moment of inertia of the motor armature is 0.01 ft-lb-sec^2.

How long will it take the motor to accelerate a flywheel from rest to 500 RPM? Assume the flywheel is coupled directly to the motor, and neglect friction due to bearings or windage. The flywheel is a 12 inch diameter steel cylinder, 6 inches long. The density of steel is 0.283 lbs/in^3.

SOLUTION: Calculate $I_{x\text{ flywheel}}$ as follows:

$$\text{Flywheel weight }(W) = \text{Density }(\gamma) \times \text{Volume}$$

$$= \gamma \frac{\pi D^2}{4} L$$

$$= (0.283) \frac{(3.14)(12)^2}{4} (6) = 192 \text{ lb.}$$

From Fig. 6-10(a):

$$I_{x\text{ flywheel}} = 0.00388\, W D^2$$
$$= 0.00388(192)(1)^2$$
$$= 0.745 \text{ lb-ft-sec}^2$$

It is important to remember that the motor must use some of its available torque to accelerate its own armature, as well as the load, up to 500 rpm.

Therefore the total moment of inertia seen by the motor is:

$$I_{x\,total} = I_{x\,armature} + I_{x\,flywheel}$$
$$= 0.01 + 0.745$$
$$= 0.755 \text{ lb-ft-sec}^2$$

Use Eq. (6-17) to calculate time T required to accelerate $I_{x\,total}$ to 500 rpm with constant torque T of 240 in.-lb (20 ft-lb):

$$t = 0.105 \frac{I_x}{T} N$$

$$= 0.105 \frac{(0.755)}{(20)} (500) = 1.98 \text{ seconds.}$$

EXAMPLE 6-3

A gear reducer with a 10:1 reduction ratio is coupled between the motor and flywheel of Example 6-2. How long does it take the same motor putting out the same torque to come up to maximum motor speed of 500 rpm?

If the motor was capable of applying a constant 20 ft-lb of torque to accelerate the flywheel to 500 rpm, how long would it take and how fast would the motor be running?

What constant torque output would be required to bring the flywheel up to 500 rpm in 2 seconds—the time required in the example where no gear train was between the motor and the flywheel?

Assume that there are no friction losses in the gear reducer or flywheel bearings, and neglect the inertia of the gears.

SOLUTION: The gear reducer has a gear ratio G of 10 (diameter of the driven or output gear divided by the diameter of the driver or input gear), as shown in Eq. (6-19).

The moment of inertia of the flywheel "reflected" back to the motor shaft is then calculated from Eq. (6-20) as

$$I_{flywheel\,refl} = I_{x\,ref.} = \frac{I_{flywheel}}{G^2}$$

$$= \frac{(0.745) \text{ lb-ft-sec}^2}{(10)^2}$$

$$= 0.00745 \text{ lb-ft-sec}^2$$

Where $I_{flywheel}$ was calculated previously in Example 6-2.

The total moment of inertia seen by the motor will be the sum of the inertia of the motor armature plus the reflected inertia of the flywheel:

$$I_{x\text{ total}} = I_{x\text{ armature}} + I_{x\text{ ref}} = 0.01 + 0.00745 \text{ lb-ft-sec}^2$$

$$= 0.01745 \text{ lb-ft-sec}^2$$

Use Eq. (6-17) to calculate time t required to accelerate the flywheel:

$$t = 0.105 \, \frac{I_{x\text{ total}}}{T} \, N = 0.105 \, \frac{(0.01745)}{20} \, 500 = 0.046 \text{ sec.}$$

Note, however, that even though the motor is running at 500 rpm, the flywheel is running at 50 rpm by virtue of the gear reducer.

If it were possible to continue to apply a constant 20 ft-lb torque to the gear reducer, and accelerate the flywheel up to 500 rpm, the motor would have to run up to 5000 rpm. We can use Eq. (6-17) directly to calculate the time as before, but since the inertia I_{total} is referenced to the motor shaft, the value for N will be 5000 rpm:

$$t = 0.105 \, \frac{I_{\text{total}}}{T} \, N = 0.105 \, \frac{0.01745}{20} \, 5000 = 0.46 \text{ sec.}$$

If it was required to accelerate the flywheel to 500 rpm (motor to 5000 rpm) in the same time as calculated in Example 6-2 (i.e., 2 seconds) the torque required to do so is calculated from Eq. (6-17):

$$T = .105 \, \frac{I_{\text{total}}}{t} \, N = 0.105 \, \frac{0.01745}{2} \, 5000 = 4.6 \text{ ft-lb.}$$

The following table summarizes the peak power required to accomplish each of the above accelerations of the load:

Motor torque output	Max. motor speed	Final flywheel speed	Peak motor output power	Time to speed
20 ft-lb	500 rpm	50 rpm	1.9 hp	46 m sec.
20 ft-lb	5000 rpm	500 rpm	19 hp	0.46 sec.
4.6 ft-lb	5000 rpm	500 rpm	4.4 hp	2 sec.

Note that the peak power requirement to accelerate the flywheel from 0 to 500 rpm without a gear reduction (Example 6-2) required only 1.9 hp (20 ft-lb at 500 rpm). Compare this with the 4.4 hp required to accelerate the flywheel to 500 rpm in the same time period (i.e., 2 seconds). The reason is that the armature inertia has become the predominant inertia as a result of the gear reducer, and the requirement to accelerate the motor to 5000 rpm instead of 500 rpm is not offset by the reduction in "reflected" flywheel inertia.

This example illustrates a common occurrence in machinery design—motors driving loads through force-multiplying transmissions such as gear reducers or lead screws. Because the transmission reduces the polar moment of inertia of the load "reflected" back to the motor shaft, it is not uncommon to find the motor armature inertia to be the predominant inertia which must be accelerated.

Notice that if the gear ratio G is greater than 1, i.e. input torque is multiplied at the output shaft, then the reflected moment of inertia is less than that of the load I_x itself. This means that a motor capable of delivering less torque can accelerate the load, or, conversely, that a small motor can "hold" a large load at rest.

Fig. 6-11 shows reflected moments of inertia for the three cases most common in machinery design: a load driven through a gear train, a load driven through a lead screw, and a load driven on a belt or conveyor.

6.3.3.   Machine Loads—Metal Forming and Metal Removal

The previous two sections discussed frictional loads and inertial loads, both of which may be considered machine loads for fulfilling material transfer requirements. Often the requirement just to move the workpiece constitutes the major load on the machine actuator, particularly in high-speed machine. The major requirement for application of forces at a work station fall into the category of material forming or removal. In a machine tool, this operation might be the primary machine function. In an assembly machine or packaging machine, on-line stamping or forming of various materials might be a secondary operation supporting the main function of the machine.

Forces Required to Bend Metal.   Fig. 6-12 shows the three common configurations for bending material: the edge bend, V-bend, and channel bend. To deform a metal, it must be stressed in its plastic region, with an average stress somewhere between the yield stress and the ultimate tensile stress. To assure that all strain hardening materials can be fully bent, bending forces are usually calculated based on material ultimate tensile strength.

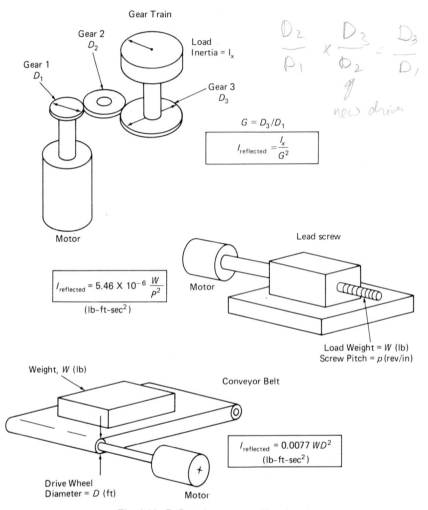

$$\frac{D_2}{D_1} \times \frac{D_3}{D_2} = \frac{D_3}{D_1}$$

new drive

$$G = D_3/D_1$$

$$I_{\text{reflected}} = \frac{I_x}{G^2}$$

$$I_{\text{reflected}} = 5.46 \times 10^{-6} \frac{W}{p^2}$$

$$(\text{lb-ft-sec}^2)$$

$$I_{\text{reflected}} = 0.0077 \, WD^2$$

$$(\text{lb-ft-sec}^2)$$

Fig. 6-11. Reflected moments of inertia.

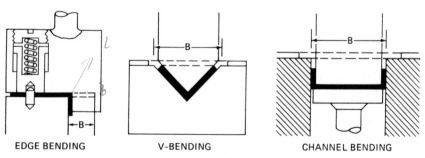

EDGE BENDING   V-BENDING   CHANNEL BENDING

Fig. 6-12. Material bending configurations.

For edge bends, the bending force P (lb) is:

$$P = \frac{L\delta^2}{2B}\,\sigma_t; \quad [lbs] \tag{6-21}$$

for symmetrical V-bends:

$$P = \frac{L\delta^2}{B}\,\sigma_t; \tag{6-22}$$

for channel or U-bends:

$$P = \frac{2L\delta^2}{B}\,\sigma_t; \tag{6-23}$$

where: L = Length of the bend (in.)
δ = Stock thickness (in.)
B = Width of unsupported material (in.) (see Fig. 6-12)
σ_t = Ultimate tensile strength of the material (psi) (See Table 6-5).

Forces Required to Shear Material. The discussion which follows relates to the operations of shearing, punching, blanking, piercing or cutting off of sheet material. The load required to shear a material is:

$$P = L\delta\tau_s \quad [lbs] \tag{6-24}$$

where: P = Shearing force required (lbs)
L = Total length of cut (in.)
δ = Material thickness (in.)
τ_s = Shear strength of material (psi).

The shearing stroke should be as quick as possible, and the shearing edge should be sharp. Most materials shear through before the cutting edge penetrates one-half of the sheet thickness. Softer, more ductile materials may require more penetration prior to severance.

Dull tools increase required shearing force by up to 30%. The shearing actuator should therefore be sized to provide sufficient force even though the cutting edge may become dull.

Table 6-5 presents shear strengths and ultimate tensile strengths for several engineering materials.

Power Required to Remove Material. Defining the machine loading for a machine tool which actually removes material by turning, milling, or

TABLE 6-5. Material Tensile and Shear
Strength

Material	Ultimate Tensile Strength σ_t (psi)	Shear Strength σ_s (psi)
Aluminum		
24SO (annealed)	25,000	18,000
-T	68,000	41,000
61SO (annealed)	18,000	12,400
-T4	35,000	13,400
75SO (annealed)	32,000	—
-T6	76,000	46,000
Commercial bronze 90%		
annealed	41,000	32,000
hard	61,000	38,000
Brass, yellow 65%		
annealed	53,000	—
hard	74,000	43,000
Cardboard		
soft	—	4,000
hard	—	8,000
Copper		
annealed	32,000	22,000
hard	50,000	28,000
Inconel		
Cold rolled, annealed	90,000	63,000
Cold rolled, hard	156,000	83,000
Leather		
soft	—	8,600
hard	—	14,200
Lead	2,400	1,800
Magnesium, hard rolled	37,000	20,000
Nickel		
cold rolled	63,000	52,000
Paper		
soft	—	4,000
hard	—	5,000
Steel, cold rolled, hard	80,000	61,000
Stainless steel, 302	74,000	57,000

drilling is a common design requirement. In a production machine, versatility
is not usually a high priority design constraint, so the machining operation
will generally be single speed and feed as required to make the optimum cut
on the workpiece at maximum production rate. (Contrast this design
requirement to that of a toolroom lathe, where the number one design

problem is that of designing a transmission system which will give all required spindle speeds.)

The technique for sizing spindle drive motors most often employed is that using the "unit power" approach. Metcut Research Associates, Inc. of Cincinnati, Ohio have compiled an extensive *Machining Data Handbook* which tabulates the proper machining speeds (of tool relative to work) and feeds (velocity of tool at right angle to cutting speed) for numerous different metals of various hardnesses, for several tool types, and for two depth of cuts (i.e. light and heavy). This data is listed for the 35 different machining processes shown in Table 6-6.

The first step in sizing the system is to specify the material to be machined and its Brinell hardness (BHN), the type of tool to be used (high speed steel, cast alloy, carbide tip), the depth of cut desired (rough or finish cut) and of course, the machining process being considered. From the appropriate table, one then identifies the proper cutting speed V_c, in feet per minute, and the proper feed rate f_m, in inches per minute.

Using the proper cutting speed V_c, the designer then calculates spindle speed (rpm) for a given cutter (or work) diameter by:

$$\text{rpm} = 3.82 \frac{V_c}{D} \qquad (6\text{-}25)$$

TABLE 6-6. Machining Operations Covered in The *Machining Data Handbook*, Metcut Research Associates, Inc. Cincinnati, Ohio.

1. Turning, single point and box tools	18. Trepanning
2. Turning, cutoff and form tools	19. Reaming
3. Turning with ceramic tools	20. Tapping
4. Boring	21. Gear hobbing
5. Counterboring and spotfacing	22. Gear cutting
6. Planing	23. Gear shaping
7. Broaching	24. Gear shaving
8. Face milling	25. Surface grinding
9. Slab milling	26. Cylindrical Grinding
10. End milling	27. Internal grinding
11. Hollow milling	28. Centerless grinding
12. Thread milling	29. Centerless grinding work traverse rates
13. Circular sawing and metal slitting	30. Gear grinding
14. Circular sawing, carbide	31. Thread grinding
15. Power hack sawing and power band sawing	32. Abrasive cutoff
	33. Honing
16. Drilling	34. Burnishing
17. Gun drilling and reaming	35. Cold form tapping

where: V_c = Cutting speed in ft/min as taken from the appropriate table
D = Diameter of work in inches for turning operation
= Diameter of cutter in inches for milling operation
= Diameter of drill in inches for drilling operation.

Note that there is a certain amount of design latitude in the selection of tool (or work) diameter and spindle speed. A larger cutter would be run at a slower spindle speed, so for the same power output, the motor torque requirement would be higher. Choice of spindle rpm may be somewhat limited, however, by the cutter feed rate, because the feed is often driven by a geared down power take-off from the spindle drive motor. The transformation from rotary to linear motion is usually made with a lead screw which will have limited practical pitch. This means that for every revolution of the spindle, the tool feed will advance by a constant linear distance appropriately called the feed per revolution f_r, related to the tool feed rate f_m by

$$f_m = f_r \cdot \text{rpm} \qquad (6\text{-}26)$$

where: f_m = Tool feed rate (in./min)
f_r = Tool feed rate (in./rev).

Knowing both cutting speed V_c and feed rate f_m allows the designer to calculate the rate of material removal Q as follows:

$$Q_t = 12 df_m V_c / \text{rpm} \qquad \text{for turning operation} \qquad (6\text{-}27a)$$

$$Q_m = wdf_m \qquad \text{for milling operation} \qquad (6\text{-}27b)$$

$$Q_d = \pi D^2 f_m / 4 \qquad \text{for drilling operation} \qquad (6\text{-}27c)$$

where: Q = Rate of metal removal (in.^{3}/min)
d = Depth of cut as originally postulated (in.)
w = width of cut by a miller (in.)
V_c = Cutting speed (ft/min), from table
f_m = Feed rate (in./min)
D = Drill diameter.

Having calculated the proper rate of material removal Q, the designer then goes to the table listing average unit power requirements for the type of operation under consideration. Table 6-7 presents some of this information from the *Machining Data Handbook*. Unit power P is in units of horsepower per cubic inch per minute, and is corrected for a dull tool (considered a worst case) and 80% spindle transmission efficiency. If the spindle efficiency is known to be different, then the unit power values in the table should be corrected. The spindle drive power requirement can then be calculated as:

$$HP_t = Q_t \cdot P_t \qquad \text{for turning} \qquad (6\text{-}28a)$$

$$HP_m = Q_m \cdot P_m \qquad \text{for milling} \qquad (6\text{-}28b)$$

$$HP_d = Q_d \cdot P_d \qquad \text{for drilling} \qquad (6\text{-}28c)$$

where: HP = Shaft horsepower required at drive motor for appropriate machining operation

Q = Rate of metal removal from Eq. (6-27) (in.³/min)

P = Average unit power requirement from Table 6-7 (hp/in.³/min).

Additional relevant design data available in the *Machining Data Handbook* includes a chart to identify the need for and proper type of cutting fluids, and a chart describing the proper cutting tool geometry for the machining process selected and its associated cutting speed and feed.

An alternative method for calculating torque and thrust forces due to drilling has been developed by Shaw and Oxford (1957). Here, the cutting forces are calculated directly from empirical equations, and machine power requirements must be calculated by the use of Eqs. (6-1) and (6-2). For drilling into solid metal:

$$M = K f_r^{0.8} d^{1.8} \frac{1 - (c/d)}{1 + (c/d)^{0.2}} + 3.2 \left(\frac{c}{d}\right)^{1.8} \qquad (6\text{-}29)$$

$$T = 2K f_r^{0.8} d^{1.8} \frac{1 - (c/d)}{1 + (c/d)^{0.2}} + 2.2 \left(\frac{c}{d}\right)^{0.8} + Kd^2 \left(\frac{c}{d}\right)^2 \qquad (6\text{-}30)$$

where: M = Required torque (in.-lb)

T = Required thrust force (lb)

K = Material work constant from Table 6-8

f_r = Feed per drill revolution (in./rev)

d = Drill diameter (in.)

c = Chisel edge length (in.) (see Fig. 6-13).

To calculate torques and thrust forces required to enlarge an existing hole by counterboring, core drilling, countersinking, or reaming the following

TABLE 6-7. Average Unit Power Requirements

MATERIAL	HARDNESS Bhn	UNIT POWER* hp/in³/min					
		TURNING P_t HSS AND CARBIDE TOOLS (feed .005-.020 ipr)		DRILLING P_d HSS DRILLS (feed .002-.008 ipr)		MILLING P_m HSS AND CARBIDE TOOLS (feed .005-.012 ipt)	
		Sharp Tool	Dull Tool	Sharp Tool	Dull Tool	Sharp Tool	Dull Tool
STEELS, WROUGHT AND CAST Plain Carbon Alloy Steels Tool Steels	85-200	1.1	1.4	1.0	1.3	1.1	1.4
	35-40 R_c	1.4	1.7	1.4	1.7	1.5	1.9
	40-50 R_c	1.5	1.9	1.7	2.1	1.8	2.2
	50-55 R_c	2.0	2.5	2.1	2.6	2.1	2.6
	55-58 R_c	3.4	4.2	2.6	3.2⁺	2.6	3.2
CAST IRONS Gray, Ductile and Malleable	110-190	0.7	0.9	1.0	1.2	0.6	0.8
	190-320	1.4	1.7	1.6	2.0	1.1	1.4
STAINLESS STEELS, WROUGHT AND CAST Ferritic, Austenitic and Martensitic	135-275	1.3	1.6	1.1	1.4	1.4	1.7
	30-45 R_c	1.4	1.7	1.2	1.5	1.5	1.9
PRECIPITATION HARDENING STAINLESS STEELS	150-450	1.4	1.7	1.2	1.5	1.5	1.9

Material	Hardness						
TITANIUM	250-375	1.2	1.5	1.1	1.4	1.1	1.4
HIGH TEMPERATURE ALLOYS Nickel and Cobalt Base	200-360	2.5	3.1	2.0	2.5	2.0	2.5
Iron Base	180-320	1.6	2.0	1.2	1.5	1.6	2.0
REFRACTORY ALLOYS Tungsten	321	2.8	3.5	2.6	3.3†	2.9	3.6
Molybdenum	229	2.0	2.5	1.6	2.0	1.6	2.0
Columbium	217	1.7	2.1	1.4	1.7	1.5	1.9
Tantalum	210	2.8	3.5	2.1	2.6	2.0	2.5
NICKEL ALLOYS	80-360	2.0	2.5	1.8	2.2	1.9	2.4
ALUMINUM ALLOYS	30-150 500 kg	0.25	0.3	0.16	0.2	0.32	0.4
MAGNESIUM ALLOYS	40-90 500 kg	0.16	0.2	0.16	0.2	0.16	0.2
COPPER	80 R_B	1.0	1.2	0.9	1.1	1.0	1.2
COPPER ALLOYS	10-80 R_B 80-100 R_B	0.64 1.0	0.8 1.2	0.48 0.8	0.6 1.0	0.64 1.0	0.8 1.2

*Power requirements at spindle drive motor, corrected for 80% spindle drive efficiency.

† Carbide

Reproduced from the *Machining Data Handbook*, 3rd Edition, by permission of the Machinability Data Center.

©1980 by Metcut Research Associates, Inc.

TABLE 6-8. Material Constant K for Eq. (6-29), (6-30).

Work Material	K
Steel, 200 BHN	24,000
Steel, 300 BHN	31,000
Steel, 400 BHN	34,000
Most aluminum alloys	7,000
Most magnesium alloys	4,000
Most brasses	14,000
Leaded brass	7,000

equations are used:

$$M = KRf_r^{0.8}d^{0.8}\,\frac{1 - (c/d)^2}{1 + (c/d)^{0.2}} \tag{6-31}$$

$$T = 2KRf_r^{0.8}d^{0.8}\,\frac{1 - (c/d)^2}{1 + (c/d)^{0.2}} \tag{6-32}$$

where: c = Existing hole diameter (in.)
d = Largest tool diameter engaging workpiece (in.) (see Fig. 6-13)
R = Correction factor for number of flutes on drill as shown in Table 6-9.

All other variables are as defined for Eqs. (6-29) and (6-30).

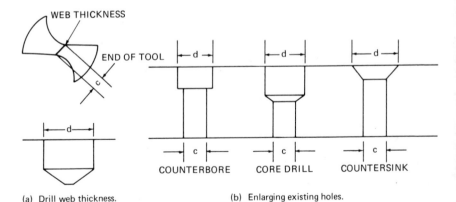

(a) Drill web thickness.   (b) Enlarging existing holes.

Fig. 6-13. Nomenclature for drilling eqs. (6-29) to (6-32).

TABLE 6-9.   Correction Factor R for
Eq. (6-31), (6-32).

Flutes	Factor R	Flutes	Factor R
1	0.87	8	1.32
2	1.00	10	1.38
3	1.09	12	1.44
4	1.15	14	1.48
5	1.20	16	1.52
6	1.25	20	1.60

The authors recommend that 30% be allowed as additional torque due to a dull drill. They also point out that water based cutting fluids have little effect on torque required, but that oil based fluids will reduce cutting torque by as much as 15%.

6.3.4.   Forces Required to Dispense a Viscous Material

Adhesives, caulking, or made-in-place gaskets are commonly applied by a syringelike device as shown in Fig. (6-14). The actuator which applies force F to the syringe plunger may be a pneumatic cylinder, a hydraulic cylinder, or an electric linear actuator (motor operating through a lead screw).

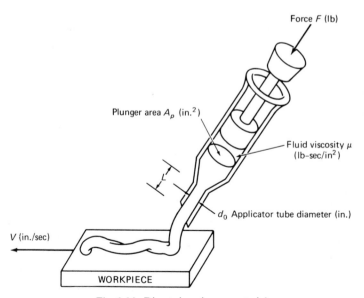

Fig. 6-14. Dispensing viscous material.

The small tube dimension d_0 is selected to lay down the size of bead required by the assembly. If the required rate of bead feed-out is V inches per second, then a steady force F must be applied to the plunger as follows:

$$F = \frac{32 \, \mu L V A_p}{d_0^2} \qquad (6\text{-}33)$$

where: F = Required force, lb
μ = Material viscosity, lb-sec/in.^{2} or Reyn
L = Length of applicator tube, in.
V = Bead feed-out velocity, in./sec
A_p = Cross sectional area of plunger, in.^{2}
d_0 = Applicator tube diameter, in.

A common problem with all adhesive systems applied in this manner is that the adhesive viscosity increases with time as the adhesive starts to set up. Therefore the plunger actuator must be sized so as to be able to apply the force seen at maximum material viscosity. Also, viscosity is very temperature dependent, and force requirements on a particularly cold day may be high. Heat can be applied to the adhesive to facilitate application flow, but in the case of epoxy adhesives, the increased temperature also decreases pot life. As a rule of thumb, epoxy adhesive cure time is cut in half for every $10°C$ ($18°F$) rise in temperature.

The best technique for applying two- or three-component catalyzed adhesives (e.g., epoxy) is to mix the components in a head just prior to application to an assembly. In this way, catalyzed material does not dwell in the machine applicator long enough to set up and cure.

Equation (6-33) can be applied to dispensing of any viscous liquids, even to filling through a tube of length L.

6.3.5. Calculating Actuator Requirements from Machine Loads and Losses

Design of the machine power train begins at the workpiece and progresses back to the actuator. First, machine loads are determined based on the specific operation being performed. Sections 6.3.1–6.3.4 discuss methods of calculating machine loads in terms of forces and torques. These forces and torques and the speed at which they must be applied (i.e., power requirement) must be transformed into torques and speeds at the actuator shaft. Since the transmission system (see Fig. 6-4) may provide a mechanical advantage from actuator to workpiece, the torque/speed requirement at the workpiece may be (and usually is) different from the torque/speed requirement at the actuator, even though the power requirement, assuming a 100% efficient transmission, is the same at both points.

The next step, then, after determining the machine load is to translate that load into actuator requirements. That requirement will be a torque and speed (or force and velocity) at the actuator shaft.

To be added to the actuator torque requirement are all of the friction torques which the transmission system contains and which must be overcome by the actuator as it drives the machine.

If the actuator is being used to accelerate and decelerate a mass, then the inertial load requirement must be added to the actuator capability. The moments of inertia to be considered include the moment of inertia of the load *reflected back to the actuator*, the moments of inertia of each component of the transmission system *reflected back to the actuator*, and finally the moment of inertia of the actuator itself.

6.4. ELECTRIC MOTOR PRINCIPLES

Electric motors offer the widest variety of power/speed/controllability capabilities in an off-the-shelf package of all competing types of actuator. It is no wonder, then, that electric motors comprise the most common type of automatic machine prime mover.

6.4.1. General Considerations

The machine designer's concern with electric motors as the primary drive actuator, or prime mover, in a special machine is twofold:

1. Selection of a motor which meets machine requirements,
2. Control of the selected motor.

These two design steps cannot be considered independently of each other. The desired machine cycle and nature of the machine drive system will dictate motor power output requirements versus time (also called *duty cycle*). The machine control system must implement the required machine cycle, hence a motor must be selected which can not only deliver the required power as demanded, but which can respond to the machine controls.

Electric Motor Ratings and Standards.   Electric motor rating methods, operating characteristics and application methods have been standardized by the National Electrical Manufacturers Association (NEMA). The NEMA publication, "Motors and Generators, MG1" provides comprehensive documentation of these standards and rating systems.

One very practical standard of use to machinery designers is that of standard motor frame sizes and their associated mounting dimensions. Various standard motor output capacities (rated horsepower) are assigned to standard size frames, making it possible to substitute different manufacturer's motors on a machine which has already been built. Also, the standardization allows the designer to lay out the machine without first having to obtain detailed motor dimension data from one or more manufacturers. Appendix I contains Tables showing frame size assignments for the very common three phase squirrel cage induction motor, and associated frame dimensions for both foot mounted and face mounted motors.

A second important NEMA specification is a classification of motor enclosures and a definition of the protection which each must provide. Table 6-10 provides a brief description of enclosure types and associated protection.

Electric motors can fail in two modes: mechanical (bearings) or electrical

TABLE 6-10.   NEMA Motor Enclosures.

Enclosure Types	Enclosure Requirements
Open	
Dripproof	Operate with dripping liquids up to 15 deg from vertical
Splashproof	Operate with splashing liquids up to 100 deg from vertical
Guarded	Guarded by limited size openings (less than $\frac{3}{4}$ in.)
Semiguarded	Only top half of motor guarded
Dripproof fully guarded	Dripproof motor with limited size openings
Externally ventilated	Ventilated with separate motor-driven blower; can have other types of protection
Pipe ventilated	Openings accept inlet ducts or pipe for air cooling
Weather protected Type 1	Ventilating passages minimize entrance of rain, snow, and airborne particles; passages are less than $\frac{3}{4}$ in. in diam.
Weather protected Type 2	Motors have, in addition to Type 1, passages to discharge high-velocity particles blown into the motor
Totally Enclosed	
Nonventilated (TENV)	Not equipped for external cooling
Fan cooled (TEFC)	Cooled by external integral fan
Explosionproof	Withstands internal gas explosion; prevents ignition of external gas
Dust-ignitionproof	Excludes ignitable amounts of dust and amounts of dust that would degrade performance
Waterproof	Excludes leakage except around shaft
Pipe ventilated	Openings accept inlet ducts or pipe for air cooling
Water cooled	Cooled by circulating water
Water-and-air cooled	Cooled by water cooled air
Guarded TEFC	Fan cooled and guarded by limited-size openings
Encapsulated	Has resin-filled windings for severe operating conditions

TABLE 6-11. NEMA Standard Temperature Rise.

	Insulation class		
	B	F	H
Open or TEFC motors without SF; rise at rated load	80	105	125
All motors with 1.15 SF; rise at 115% load	90	115	—

(winding insulation). Electrical or winding failure is the most prevalent mode, and the greatest cause of insulation failure is thermal overload. Winding temperature versus winding life curves have been generated from experimental data and correlated by a method known as IEEE #510 life curves. A very important rule of thumb which can be derived from these curves is:

Motor winding nominal life is reduced by a factor of one-half for each 10°C hotter that the motor runs.

Using a nominal motor life of 20,000 hours, NEMA has specified standard temperature rises allowed for various classes of insulation (see Table 6-11). The allowable temperature rise assumes an ambient temperature of 40°C and allows an additional 10°C hot spot contribution. This means that the actual winding temperature which would result in a motor life of 20,000 hours is the sum of the allowable rise shown in Table 6-11 plus 50°C.

NEMA has further specified a motor service factor which indicates how much over rated load a motor can run without exceeding the temperature rise standard. Most standard motors under 200 hp must have a service factor of at least 1.15, which means that these motors must be able to be run at 115% of rated or nominal (nameplate) horsepower *without* exceeding the standard temperature rise for the insulation type used in that motor. Motors over 200 hp have no standard service factor requirement.

Motor Types. Fig. 6-15 shows a breakdown of motor types by electrical configuration. The two basic types of motors are DC and AC motors, classified by the form of electrical power required at the motor terminals. Since the predominant form of electricity available in the world today is AC, even DC motors require an AC/DC power converter of some type to supply input power.

Historically, DC motors were selected for applications in machinery requiring adjustable speed drives because speed control over a relatively wide range (30:1) is easily accomplished by varying the magnitude of terminal supply voltage to either the motor armature or the field windings. Modern

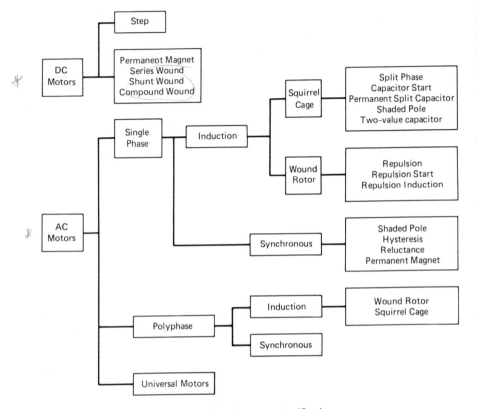

Fig. 6-15. Electric motor classifications.

electronic AC motor speed controls make it possible to use AC motors in variable speed applications, so to a certain extent, speed control is no longer the dominant selection criterion favoring DC motors.

The selection of a motor type is often dictated by the specific requirements of the special machine, in that only one type will meet the performance criteria. The choice of motor type is not always clear, however, in cases where more than one type of motor will do the job. The following generalizations concerning motor selection for special machine applications reflect composite industry practice, but *may not be applicable* to a specific application with unique constraints. The designer should therefore be wary of generalizations and consider each application as a separate case.

- For a constant speed drive with light to moderate load fluctuations and a requirement for high reliability and minimal maintenance, use a three-phase squirrel cage induction motor.

- Always use three-phase AC motors (rather than single-phase) if possible for maximum efficiency and minimal vibration.
- For rapidly changing variable speed drive with constant torque required over a range of speeds, use a permanent magnet DC motor for applications less than one horsepower. Use a wound-field DC motor (shunt or compound) for higher horsepower applications.
- For on-off driving of heavy loads (e.g., conveyors) where fast, controlled acceleration is not required, use an AC squirrel cage induction gearmotor.
- For accurate x-y positioning, use either a step motor or a permanent magnet DC servomotor. Use a step motor for low power applications, DC servomotor for higher power requirements.

The following sections discuss AC motors (Section 6.4.2), DC motors (Section 6.4.3), step motors (Section 6.4.4) and servomotors (6.4.5) in greater detail.

Motor Loads and Motor Sizing. Section 6.1 discussed methods for calculating various machine loads imposed on the machine by its operating requirements. These requirements may be stated in terms of a machine power requirement or in terms of a machine torque and speed (rpm) requirement. Clearly the drive motor will have to be rated at slightly higher horsepower than the total machine load (to account for inefficiencies in the machine itself) but horsepower alone does not match a motor to a specific machine requirement.

In order to fully predict how a particular motor will drive a particular machine, we need to have an output *torque-speed curve* for the motor, a required torque-speed characteristic for the machine itself, and a required torque versus time characteristic of the machine load.

There are three levels of torque which the motor must be capable of delivering:

1. *Breakaway Torque.* The motor must be capable of delivering sufficient torque at zero speed to get the machine started. Breakaway torque is always higher than normal running torque due to static coefficients of friction and machine imbalances. If the motor can not generate torque at least equal to the machine breakaway torque,* then the machine will not start.

2. *Accelerating Torque.* The motor must be able to deliver some excess torque over that required to just match machine load in order to bring the

---
*Throughout Section 6.4, "machine load torques" refer to torque requirements at the motor shaft, not at the machine output.

machine up to operating speed. The greater this excess torque, the faster the machine will accelerate. If the torque required by the machine load increases to equal motor output torque at a speed lower than operating speed, then the machine will never attain operating speed.

3. *Running Torque.* Once up to speed, the motor output torque must equal the machine load torque, and be capable of supplying any fluctuations in required load torque.

Only consideration of the individual motor torque–speed curve will show whether or not the motor will (a) start, (b) come up to speed, and (c) overcome fluctuations in running torque. Characteristic curves for each type of motor are discussed in the sections which follow.

The steady state or running power requirement of the motor will be a function of the running load torque as seen by the motor and of motor speed:

$$P_m = \frac{T_l \cdot N}{5{,}250} \tag{6-34}$$

where: P_m = Steady state power requirement or rated load of motor (hp)
T_l = Torque load due to machine loads and machine as seen by the motor at its shaft (ft-lb)
N = Steady state motor operating speed or rated speed (rpm).

Motors are designed for maximum efficiency at or near the motor rated load (i.e., the nominal motor characteristics specified on the motor nameplate), thus it is important to match machine load to motor rated load. Continuous operation outside ±10% of rated load usually results in the motor running at higher temperatures, which in turn reduces motor operating life.

The time required to accelerate the machine to speed, N (rpm) from rest can be calculated from Eq. (6-17) as follows:

$$t = 0.105 \frac{I_x N}{(T_m - T_l)} \tag{6-35}$$

where: t = Time in seconds to accelerate machine
T_m = Constant torque delivered by the motor during acceleration (ft-lb)
T_l = Load torque due to machine loads as seen by the motor at the motor shaft (ft-lb)
I_x = Moment of inertia of the machine load and machine power transmission system reflected back to the motor shaft *plus* the moment of inertia of the motor rotor and shaft (ft-lb-sec²)
N = Speed which motor and machine eventually reach (rpm).

Calculation of acceleration time is important because motors will put out maximum torque possible during acceleration. For both AC and DC motors, this means operation at 150% or greater of rated torque, which in turn means 150% or more of normal motor current draw. If the motor makes many starts and stops during operation, temperature rise will be significant, and motor life will be shortened accordingly. If starting time exceeds 45 seconds, the motor torque is insufficient and a larger motor should be considered.

6.4.2. AC Motors

Basic Principles—Rotating Magnetic Fields.   All AC motors, whether they be induction motors or synchronous motors, consist of a rotating part called the *rotor* and a set of stationary wire wound coils called the *stator*. Fig. 6-16 shows a typical wire-wound stator. The stator *core* is made up of a stack of individually insulated, magnetically soft laminations shaped as shown in Fig. 6-16(a). Depending on motor design, an even number of *poles* are created

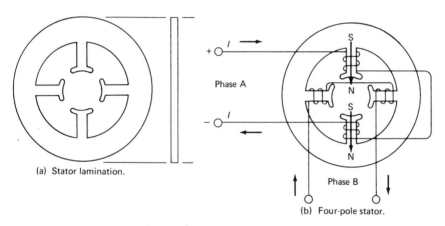

(a) Stator lamination.

(b) Four-pole stator.

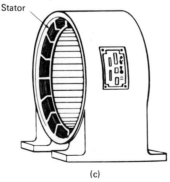

Fig. 6-16. AC motor stator design.

by winding magnet wire through the slots in the stator core. Pole faces opposite each other are wound with the same wire, so that if a steady current flowed through the wire, one pole face would be a North (N) magnetic pole and the opposite pole face would be a South (S) magnetic pole [see Figure 6-16(b)].

Assume that the other pair of poles on the stator of Fig. 6-16 are also wound with a separate wire so that they also form a N–S pair when current flows through that wire. Consider then the situation depicted in Fig. 6-17(a)

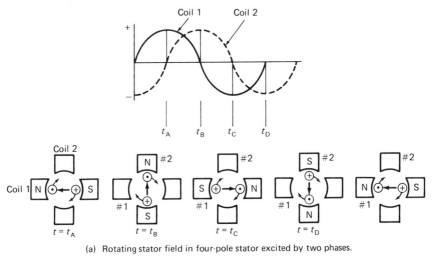

(a) Rotating stator field in four-pole stator excited by two phases.

(a)

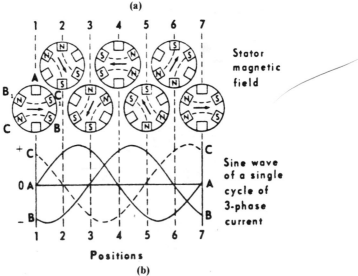

Positions

(b)

Fig. 6-17. Rotating magnetic field due to polyphase AC current in a motor stator.

where Coil 1 is energized with AC voltage as shown, and Coil 2 is energized by another AC voltage 90° out of phase with the voltage to Coil 1.

At time t_A, Coil 1 is energized with full (positive) voltage, so a N–S magnetic field is established in the stator as shown. At the same time, voltage to Coil 2 is zero, so no current flows, thus no magnetic field exists in Coil 2. At time t_B, Coil 2 has positive voltage, hence N–S field, while Coil 1 is not energized. At time t_C, Coil 1 has *negative* voltage applied, hence the field is opposite what it was at t_A. By considering the stator magnetic field through one full cycle of two electrical phases applied to the stator, we see that the magnetic field has made *one full clockwise rotation*.

Figure 6-17(b) shows a similar progression of the stator magnetic field for three coils, each powered by a separate phase of three-phase current (or voltage). Again, note the rotation of the stator magnetic field at a rate of one revolution per two poles (or *pole pair*) per phase.

Consider the stator magnetic field of Fig. 6-17(a) or 6-17(b) if all of the coils were energized simultaneously with AC power of the same phase. The resulting magnetic field would simply oscillate, not rotate, with applied power. The following conclusions may, therefore, be drawn from the previous discussion:

1. A rotating magnetic field can be created in a multipole motor stator with two or more sets of independent windings powered by out-of-phse AC power to those windings.
2. A rotating magnetic field *can not* be created by in-phase power input to stator windings.

Since single phase AC power is the most commonly available type of power, some way had to be devised to start and run single-phase AC motors. Recall from basic AC electrical theory that a capacitor or an inductor will phase shift AC current. It is therefore possible to create a quasi-two-phase power source by inserting a capacitor (or inductance) into the supply line to one of the motor pole pairs.

This quasi-two-phase power is sufficient to create a rotating stator field, but that field is not "balanced" because the capacitance (or inductance) also alters the magnitude of the shifted phase. As a result, single-phase AC motors do not run as smoothly as two- or three-phase motors. For long-life machine applications, three-phase AC motors are preferred for several reasons, one of them being low vibration while running.

It should be pointed out that a single-phase AC motor will run on single-phase power input to the stator *if* it is first brought up to speed. It is therefore common to find "capacitor start" motors which use a capacitor to create a second electrical phase for starting, but which are switched out of the circuit for running.

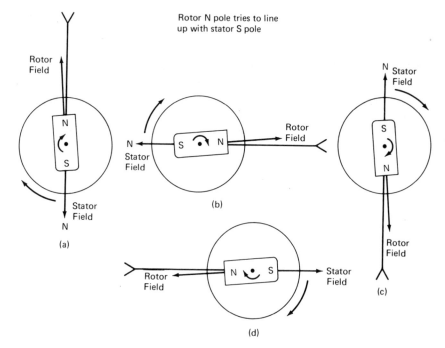

Fig. 6-18. Interaction of AC motor rotor with the rotating magnetic stator field.

It has been shown that AC motors operate by creating a rotating magnetic field in the motor stator. How does this actually drive the motor? Imagine a permanent bar magnet with its own N and S poles free to rotate about a central shaft placed into the stator of Fig. 6-16 or 6-17. Fig. 6-18 shows this situation. Recall from basic physics that like magnetic poles repel and unlike magnetic poles attract. The rotor magnet, or rotor magnetic field, will thus try to line up with the stator field as shown in Fig. 6-18. As the stator field rotates, the rotor field "chases" it, trying to line up.

The stator magnetic field rotates at a speed proportional to the frequency of the electrical power supplied to the stator, and inversely proportional to the number of stator poles (per electrical phase). The stator rotating field speed is called its *synchronous speed* and is calculated by:

$$\eta_s = 120f/p \qquad (6\text{-}36)$$

where: η_s = Motor synchronous speed (rpm)
$\quad\quad f$ = Electrical frequency supplied to motor (Hz)
$\quad\quad p$ = Number of stator poles per electrical phase.

Basic Principles—Rotor Fields. In order for the rotating stator magnetic field to drive a rotating shaft, or rotor, the rotor must itself have a magnetic field to interact with the stator field. The first possibility which comes to mind is to use permanent magnets on the rotor. If this is done, no external electrical power need be supplied to the rotor in order to create the rotor magnetic field. The stator rotating field will simply lock onto the rotor magnetic field, and "pull" the rotor around at synchronous speed. An AC motor in which the rotor rotates at the same speed as the stator magnetic field synchronous speed is called a *synchronous motor.*

Permanent magnets on the rotor are quite heavy and are limited in their maximum field strength. The next logical step in providing a rotor magnetic field would be to use electromagnets. Since we want the rotor field to be constant in magnitude and direction, we would use DC current supplied to the rotor through slip rings. This type motor is called a *wound-rotor* synchronous motor.

Because synchronous motors are not widely used in automatic machinery applications, they will not be discussed further. All further discussion will be limited to AC *induction motors* which are the most common industrial motors.

Like synchronous motors, the induction motor requires a rotor magnetic field in order to be driven by the rotating stator magnetic field. What makes an induction motor unique is that it utilizes the rotating stator field *to generate or induce rotor currents which in turn create rotor magnetic fields.*

Figure 6-19 shows a basic squirrel cage rotor used in induction motors. It provides low resistance conductors (usually aluminum) running approximately parallel to the rotor shaft between rotor pole faces (usually magnetically soft iron). The conductors are shorted to each other at both ends of rotor by aluminum end plates. Notice that each iron pole face is then encircled by a low-resistance current path which, indeed, serves as a winding.

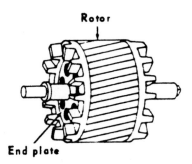

Fig. 6-19. Squirrel cage rotor for AC induction motor.

Recall from basic physics that a conductor, when moved at right angles to a magnetic field (or vice versa), induces a current in that conductor. (Faraday's Law of Induction). As the rotating stator magnetic field lines "cut" the squirrel cage conductor bars, current loops are induced which flow around the iron pole faces through the high-conductivity bars and end plates. This induced current creates rotor magnetic fields perpendicular to the pole faces, which then interact with the rotating stator field.

While the rotor is stationary, the stator magnetic field moves past the conductor bars much faster than it does when the rotor comes up to speed. The higher velocity due to relative motion between rotor and stator induces higher currents, thus larger rotor magnetic fields, thus higher motor torque. *Induction motor developed torque is highest at zero and low rotor speeds.*

As the rotor speed approaches stator field synchronous speed, the relative velocity between the two decreases, thus motor torque capability drops off. If the rotor were to attain synchronous speed, no rotor currents would be induced, thus no rotor magnetic field would exist and the motor would be capable of delivering zero torque. *Induction motor developed torque is zero at synchronous speed.*

At what speed then, does an induction motor run? It will run at the speed at which the torque load applied to the motor shaft is just equalled by the torque developed by the motor. Torque developed by the motor depends on rotor magnetic field strength, which depends on induced current in the rotor, which depends on the relative velocity of stator field with respect to the rotor speed. The difference between rotor speed and synchronous speed is called *slip*. An induction motor will slip as much as is required to produce a torque to balance load torque. *No-load slip* is the slip which occurs when the motor shaft is not loaded at all. It is produced by the motor rotor bearing torques and shaft and rotor windage. As one would expect, it is usually very small. *Slip at rated load* is the slip which occurs when a load torque equal to rated torque is applied at the motor shaft.

Slip can be expressed in rpm, i.e., the difference between synchronous speed and actual speed, or as a percentage of synchronous speed. Example 6-4 demonstrates the calculation of synchronous speed and slip.

EXAMPLE 6-4

A four-pole induction motor is powered by three-phase, 60 cycle, 220 V AC and has a rated slip of 3%. What is the motor synchronous speed? At what speed does the motor run at rated load?

SOLUTION: Number of poles $p = 4$ and frequency-60 Hz, so Eq. (6-36) yields

$$\eta_s = 120f/p = 120 \, \tfrac{60}{4} = 1800 \text{ rpm.}$$

If slip is 3%, then

$$\frac{slip}{\eta_s} = 0.03 = \frac{\eta_s - \eta_{rated}}{\eta_s} \, .$$

For $\eta_s = 1800$, $\eta_{rated} = 1746$ rpm.
Note that 1750 rpm motors are very common.

A.C. Induction Motor Characteristics. Fig. 6-20 shows a typical induction motor torque–speed curve. Motor performance curves are typically drawn with torque on the horizontal axis and speed on the vertical axis. The torque produced at zero rotor speed is called *starting torque* or *locked rotor torque*. In a machine application, the motor must have a high enough locked rotor torque to overcome machine breakaway torque. If it does not have this capability, the machine will not start. The minimum accelerating torque produced by the motor, also called *pull-up torque,* is slightly less than the locked rotor torque, but once machine breakaway torque is overcome, machine load torque usually drops considerably.

Torque increases to a value equal to or slightly greater than locked rotor torque at speeds 60–80% of synchronous speed. This "knee" of the torque-speed curve is known as *breakdown torque.* Note that if the motor and

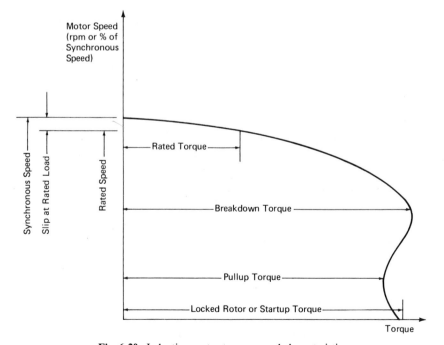

Fig. 6-20. Induction motor torque–speed characteristics.

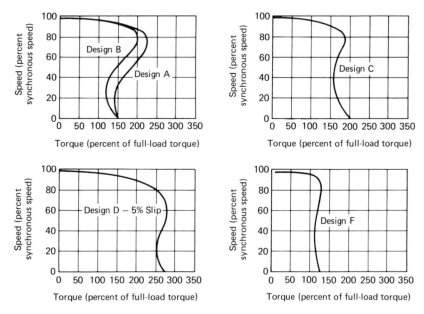

Fig. 6-21. Induction motor torque–speed characteristics for NEMA designs A, B, C, D, F.

machine are running, then the machine load increases to a torque higher than breakdown torque, the motor will quickly stall.

The motor is designed to run at a rated load (i.e., rated torque and rated speed) at or very near to maximum motor efficiency. In addition, rated temperature rise is based on the motor running at rated load, or at the service factor times rated load. Unless there are mitigating circumstances, a motor should be selected with a rated load equal to the machine load.

Figure 6-21 shows typical torque–speed curves for various NEMA classified three-phase induction motor designs. Design A motors are general purpose integral horsepower motors capable of starting machines with breakaway torques 150% of rated or running torque. Design B motors are fractional horsepower motors. Both run at approximately 3% slip. Design C motors offer high starting torque but relatively low breakdown torque. Design D motors have very high starting torques and high slip. This allows large machine load variations but the motor speed droops accordingly. Design F motors are low starting torque, low slip and can not handle significant overloads.

6.4.3. DC Motors

Basic Principles. Unlike AC motors with their characteristic rotating stator fields, DC motors have a fixed stationary magnetic field. The DC motor rotor is called the *armature,* and it consists of two or more wirewound electromagnetic poles. Fig. 6-22 depicts a simple D.C. motor. The armature magnetic field is created by passing D.C. current through the armature coil. Current flows to the armature through brushes and a segmented slip ring called a *commutator.* In the position shown in Fig. 6-22, the armature magnetic field has its north (N) pole pointing down.

The DC motor fixed field magnets may be permanent magnets or electromagnets, but they will always consist of an opposing N–S pair. For the situation shown in the figure, the armature field N pole will attempt to a'ign with the field magnet S pole, thereby creating counterclockwise rotation. The motor is developing maximum torque on the armature in the position shown, and that torque will sinusoidally drop to zero at the point where the poles line up.

However, just as the armature rotates so that is N field lines up with the S field magnet, the commutator segments rotate so that DC current now flows in the opposite direction in the armature coil. This switches the polarity of the armature magnetic field, so that now the N armature pole is lined up with the N field pole and likewise for both S poles. The like poles repel and the armature continues to rotate, trying to line its new S pole up with the field N pole. Just as it does, the armature field switches again, and the armature is again repelled as rotation continues.

As the number of armature and field poles are increased, the torque developed by the motor becomes smoother and more constant. Note that for every armature field pole pair there will be two commutator segments. As soon as one pair rotates off of the brushes, a new pair will be energized. The armature field has thus not rotated far at all before switching to a new coil.

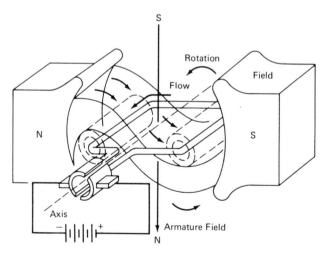

Fig. 6-22. DC motor concept. (Courtesy of *Power Transmission Design Magazine*)

The armature driving torque remains very close to its maximum value throughout the armature rotation.

There is no frequency-dependent upper limit on DC armature rpm as there is in an AC motor with its synchronous stator field speed. Armature speed in a DC motor is governed by the amount of torque which can be generated by the armature coil/field coil interaction. Torque depends on magnetic field strength which depends almost linearly on *armature current:*

$$T_m = K_T I_a \tag{6-37}$$

where: T_m = Motor torque delivered (in.-lb)
K_T = Motor torque constant (in.-lb/amp)
I_a = Armature current (amps).

The DC motor torque constant is determined by the particular motor design, and is specified for each motor by the manufacturer.

Armature current is supplied to the armature by applying a DC voltage to the motor terminals. At zero speed, the armature current I_a is

$$I_a = V/R_a \tag{6-38}$$

where: I_a = Armature current (amps)
V = Voltage applied at motor terminals (DC volts)
R_a = Armature winding resistance (ohms).

As the armature begins to rotate, however, approximately the same situation occurs as did in the squirrel cage AC induction motor. As the armature windings rotate through the fixed magnetic field of the motor field windings (or magnets) a voltage is generated (Faraday's Law of Induction) which *opposes* the applied terminal voltage. The faster the armature turns, the greater this opposing voltage becomes. It is appropriately called *back EMF* or *counter EMF* and can be written as:

$$E_B = K_B \eta \tag{6-39}$$

where: E_B = Back EMF (volts)
K_B = Back EMF constant (volts/rpm)
η = Armature speed (rpm).

Equation (6-38) is only appropriate for no armature rotation. The more general expression for armature current takes rotation into consideration as follows:

$$I_a = \frac{V - E_B}{R_a} \qquad (6\text{-}40)$$

or

$$I_a = \frac{V - K_B \eta}{R_a} \qquad (6\text{-}41)$$

$$T_m = K_T I_a$$

where: V = Terminal voltage (volts)
  K_B = Back EMF constant (volts/rpm)
  η = Motor (armature) speed (rpm)
  R_a = Armature resistance (ohms).

The ideal DC motor torque–speed relation can be obtained by combining Eqs. (6-41) and (6-37) to yield:

$$T_m = K_T I_a = \frac{K_T}{R_a} (V - K_B \eta). \qquad (6\text{-}42)$$

This relationship assumes a constant field and a linear torque–current relation which is very close to the actual case for permanent magnet DC motors and servomotors. One additional motor characteristic of concern is the internal friction of the motor. Often this friction is linear with speed and is expressed:

$$T_v = K_v \eta \qquad (6\text{-}43)$$

where: T_v = Internal viscous friction (in.-lb)
  K_v = Viscous friction constant (in.-lb/rpm)
  η = Motor speed (rpm).

If viscous friction is significant, it must be subtracted from the ideal torque expression of Eq. (6-42) to yield net developed torque. Example 6-5 illustrates the development of a torque–speed curve from motor constants given by the manufacturer's data sheet.

Shunt motor speed control by field voltage (current) control is possible because field strength affects armature back EMF. Reducing field voltage reduces back EMF generated in the motor armature, so that armature current, torque, and speed *increase*. Field control is not effective therefore in reducing motor speed to zero.

EXAMPLE 6-5

A DC permanent magnet servomotor has the following data sheet specifications:

> Torque Constant $K_T = 7.5$ inch-pounds/amp
> Back EMF Constant $K_B = 0.09$ volts/rpm
> Viscous Friction Constant $K_V = 0.105$ inch-pounds/rpm
> Armature Resistance $R_A = 0.9$ ohms.

For an applied terminal voltage of 60 volts DC, what is the maximum (no-load) motor speed? What is the maximum (stall) torque capability? What is the inrush current (armature current at zero speed)? Draw the theoretical speed–torque curve.

$T_{net} = T_m \cdot T_v$
$= 0$

SOLUTION: At no-load speed, the motor torque T_M as given by Eq. (6-42) will just equal the internal motor friction T_v as given by Eq. (6-43):

$$\frac{K_T}{R_A}(V - K_B \eta) = K_v \eta$$

Solving for motor speed η:

$$\eta = \frac{K_T V}{(K_v R_A + K_T K_B)} = \frac{(7.5)(60)}{(0.105)(0.9) + (7.5)(0.09)} = 585 \text{ rpm.}$$

Maximum torque will be generated at zero motor speed or *stall*. Eq. (6-42) is used, with speed η set to zero:

$$T_M = \frac{K_T V}{R_A} = \frac{(7.5)(60)}{(0.9)} = 500 \text{ in.-lb.}$$

Inrush current is maximum at zero motor speed because back EMF is zero, hence current is given by Eq. (6-38):

$$I_a = \frac{V}{R_A} = \frac{(60)}{(0.9)} = 66.7 \text{ amps.}$$

Typical torque-speed curves plot the motor torque delivered on the abscissa and motor speed on the ordinate. Note that torque is motor torque *net* of internal friction losses, so the plot will be of the characteristic equation:

$$T_{net} = T_M - T_v = \frac{K_T V}{R_A} - \left(\frac{K_T K_B}{R_A} + K_v\right)\eta$$

$$= 500 - 0.855\eta.$$

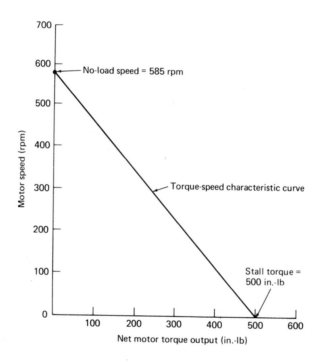

Series, Shunt, and Compound DC Motors. The preceding discussion and analysis of ideal DC motors predicts performance rather well for DC motors with permanent magnet fields. DC motors also have wirewound field coils and are often classified in terms of how the field winding is connected with respect to the armature. Fig. 6-23(a) shows the circuit diagram of a DC *shunt* motor. The field winding is wired in parallel with the armature winding. Fig. 6-23(b) shows a D.C. *series* motor wherein the field is wired in series with the armature. Fig. 6-23(c) shows the circuit for a combination shunt–series field connection known as a *compound* motor.

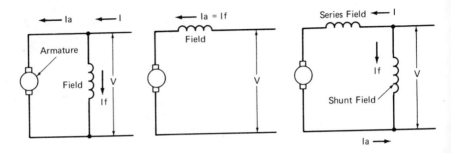

(a) Shunt-wound motor field is wired in parallel with armature.

(b) Series-wound motor field is wired in series with armature.

(c) Compound-wound motor fields are wired in series *and* in parallel with armature.

Fig. 6-23. Circuit diagrams for wound-field DC motors.

In a shunt motor, the field current, hence field strength, depends only on the resistance of the field winding. In this sense the shunt motor is like a permanent magnet motor except that field current increases and decreases with "applied terminal voltage." Figure 6-24(a) shows the characteristic torque–speed curve and current–speed curve for a DC shunt motor.

The speed of a shunt motor can be controlled by varying terminal voltage, thus armature current in much the same way as one controls a permanent magnet DC motor. The motor speed is controlled by varying full power to the motor.

If the shunt motor field is controlled independently of armature voltage, then motor speed can be controlled over a 3 : 1 range by varying only the field voltage (thus current). Since field current is typically 10–20% of armature current, motor speed control is obtained by control of relatively low power levels.

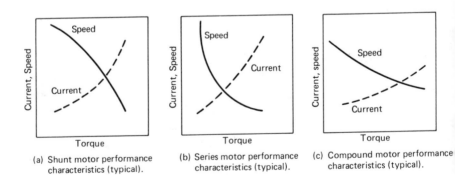

(a) Shunt motor performance characteristics (typical).

(b) Series motor performance characteristics (typical).

(c) Compound motor performance characteristics (typical).

Fig. 6-24. Torque–Speed curves for wound-field DC motors.

The advantage of DC shunt motors is that sensitivity of motor speed to changes in torque is relatively small. However, if the armature is stalled, armature current draw can be very high.

In the DC series motor [Fig. 6-23(b)], field current and armature current are equal. Maximum current flows through the armature and field at zero speed (back EMF = 0), hence high torque is developed on startup. Primary applications of DC series motors are for high starting torque requirements such as locomotives, cranes, elevators, or hoists.

Since both field current and armature current vary with armature speed, there is a pronounced effect of speed variations on torque. A series-wound motor is more sensitive to changes in load than are the other types of wound field motors. Fig. 6-24(b) shows the typical torque–speed and torque–current curves for series-connected motors.

An important advantage of series-wound motors is that armature current is limited by the field resistance at stall torque. A serious disadvantage is that the motor will run away (i.e., overspeed) if the load is removed. Also, the direction of rotation of the series motor can not be reversed by simply reversing applied voltage. A separate set of field coils connected to create a magnetic field opposite that of the series field must be switched into the circuit.

The torque–speed characteristics of a compound motor can be tailored by design of relative shunt field/series field strength. Fig. 6-24(c) shows a typical compound motor curve. Fig. 6-25 shows how the torque–speed curve can be

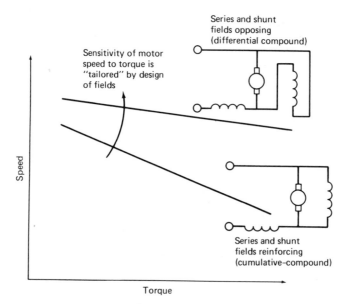

Fig. 6-25. Compound DC motor characteristics varied by field winding design.

shifted by changing the relative directions of the series and shunt fields. A *differential compound* motor has the series and shunt fields opposing each other. This results in a flat torque–speed curve. A *cumulative compound* motor has the series and shunt fields reinforcing each other for a more pronounced torque–speed slope.

Compound motors offer high starting torque and can be tailored for relatively constant speed over a wide range of torque output. The main disadvantage is that the current through both field windings must be reversed relative to the armature current in order to reverse motor direction.

6.4.4. Step Motors

Basic Principles. The motors considered in previous sections were continuous rotation motors, that is, when voltage (AC or DC) is supplied to the motor terminals, the motor accelerates up to rated speed, and continues to run at that speed unless turned off, or heavily loaded. Step motors are DC powered motors designed to output incremental motion. By energizing various stator windings individually or in various combinations, the rotor can be made to assume stable, repeatable angular positions. By energizing and de-energizing stator windings in the proper sequence, the step motor can be made to rotate continuously, or *slew*, at a velocity determined by the rate of stator coil power switching.

Step motor stators consist of multiphase wirewound poles not unlike AC motor stators. Step motor rotors of the permanent magnet type use permanent magnets on the rotor to create alternating N–S magnetic fields around the rotor periphery. Fig. 6-26 depicts a step motor with four stator poles and five permanent magnet rotor poles.

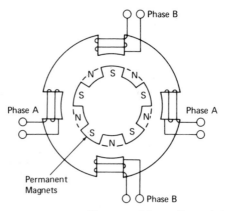

Fig. 6-26. Permanent magnet step motor. (Courtesy of *Power Transmission Design Magazine*)

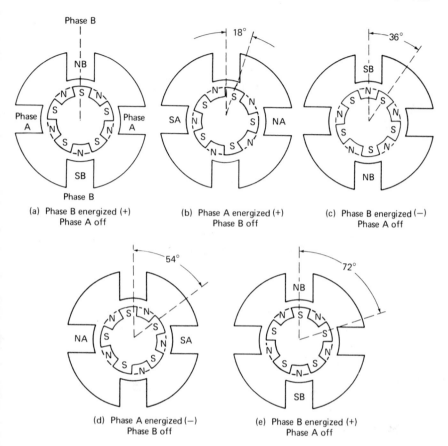

Fig. 6-27. Permanent magnet step motor stepping sequence. (Courtesy of *Power Transmission Design Magazine*)

Figure 6-27(a) shows the rotor position with phase A energized. Power to phase A is then turned off, and power supplied to phase B, creating N–S poles as shown. The rotor will rotate one *step* so that unlike magnetic poles line up [Fig. 6-27(b)]. Phase B is then de-energized and Phase A energized in the opposite sense [Fig. 6-27(c)]. The motor steps one more time.

Figure 6-28 shows a circuit for applying power to the coils by opening and closing two switches. Note that two DC power supplies are required, one positive and one negative.

In order to use only one DC power supply rather than two, the step motor stator can be *bifilar* wound. Here, two wires rather than one make up the stator coil. By switching unipolar power to one coil or the other (connected in the opposite sense to the power supply), stator polarity can be switched from N to S and vice versa. In order to make two separate windings around the same physical stator pole, thinner-gage wire must be used. Since thinner-gage wire has a higher resistance per unit length than larger-diameter wire, the L/R (inductance/resistance) ratio for each coil is less. As it turns out, the bifilar-wound step motor exhibits better dynamic characteristics (i.e., more torque at higher speed). However, for the same current per phase, the higher-resistance bifilar winding will heat faster.

Note that stepping in Fig. 6-27 is achieved by energizing adjacent phases or stator pole pairs one at a time. It is possible to energize two adjacent phases *simultaneously*, creating a stator magnetic pole whose resultant direction is halfway between stator poles. The rotor will align with this pole, therefore the capability exists for *half-stepping* the motor by simultaneously energizing adjacent poles.

Half-stepping can result in higher stepping speeds, but the designer must consult the half-stepping torque-speed curve to determine if the motor delivers sufficient torque or not. There is not a reliable analytical method for predicting half-stepping performance knowing only full-stepping performance.

A permanent magnet step motor has all of the essentials of an AC synchronous motor, and indeed, is often driven in that fashion. The rotor magnets provide the necessary rotor fields. If two-phase AC power is applied to the stator pole pairs, a rotating stator field will be established exactly as described in Section 6.4.2. The synchronous speed will be as described in Eq. (6-36).

The synchronous motor capability alone of a permanent magnet step motor would not be utilized in an automatic machine, but the dual nature (AC synchronous/DC step) might provide a useful machine feature.

Step Motor Controls. Step motors would have little practical use in automatic machine applications if mechanical switches were required to

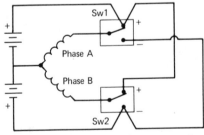

Step	CW rotation		CCW rotation	
	Sw1	Sw2	Sw1	Sw2
1	Off	+	Off	+
2	+	Off	−	Off
3	Off	−	Off	−
4	−	Off	+	Off
1	Off	+	Off	+

(a) Switching circuit with bipolar power supplies.   (b) Switching sequence for rotation of Fig. 6-27.

Fig. 6-28. Stepping circuit for motor of Fig. 6-27. (Courtesy of *Power Transmission Design Magazine*)

switch power to individual stator coils in the proper sequences. Use of electronic switching, however, renders step motors very practical and versatile.

Most step motor manufacturers also supply step motor control packages for use with their motors. It is important for the designer to base step motor selection on step motor torque–speed curves *with a given control package.* The same step motor usually exhibits different torque–speed characteristics when used with different types and models of controllers.

The simplest type of step motor controller is the *translator.* The translator provides the logic required to switch power to the appropriate stator poles so that a single DC input pulse (to the translator) results in a single angular step of the motor shaft. There are usually separate inputs for pulses to drive the motor clockwise and counterclockwise. Fig. 6-29 diagrams the performance of a step motor with translator control.

Input pulses are specified by the manufacturer in terms of voltage, minimum pulse duration, and maximum pulse rise and fall times. Translators are available to be driven by just about any standard electronic logic level signal. Translators are supplied both without and with a self-contained motor power supply. If a separate power supply must be obtained, it must be compatible with the translator as well as capable of outputting required motor phase currents.

In addition to compatibility with the particular motor, translator selection requires consideration of the following:

1. Maximum stepping speed (steps/sec);
2. Adjustable or fixed stepping speed;
3. Modular circuit board or completely packaged, self-contained unit;
4. Full step and/or half step operation.

If a computer is to drive a simple translator/step motor system, then the computer must output one pulse for each step required. If the step motor is to

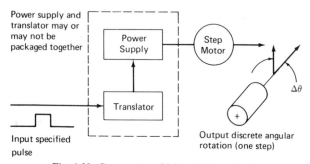

Fig. 6-29. Step motor with translator control.

drive to a specified position, then the computer must calculate how many steps are required to reach the destination (from the present position) and output the appropriate number of pulses to the translator. If acceleration, deceleration, or speed is to be controlled, then the computer must implement this by outputting the pulses at the proper rate.

The other common type of packaged step motor controller is the *indexer*. Instead of inputting one pulse for each desired motor step, the controller inputs the number of steps to be taken. Inputs can be mechanical thumbwheel settings, or computer outputs representing the number of steps required.

A complete indexer package will have:

1. Self-contained power supply,
2. Adjustable speed,
3. Adjustable acceleration/deceleration characteristics,
4. End of cycle signal output,
5. Full- or half-step operation,
6. Remote manual "run," "jog," "direction," and "clear" control.

The programming of a control computer can be simplified if an indexer is used with the step motor because the computer does not have to calculate and implement speed and acceleration profiles.

Recently, step motor control packages have become available which require the control computer only to specify the x-y coordinates desired and the control package will take care of the rest. These are not truly standard units, however, because the step motor controller must be programmed with a zero position reference and relationship between driven member position and motor steps for the particular step motor drive system in question.

Application of Step Motors. In automatic machinery design, step motors are particularly useful for (1) work positioning as in x, x-y, x-y-z tables, and (2) tool positioning. The practical power limit of electric step motors is about one horsepower. However, many step motor positioning systems are relatively low-power applications.

Motors are readily available with step sizes ranging from 90° down to 0.72°, with steps in the range of 15–1.8° being most common. Recall also that step resolutions of $\frac{1}{2}$ may be achieved with electronic half-stepping. The positioning accuracy of the step motor is 1–5% of the step angle, with 3% being a common stated accuracy.

Step motor speed is usually specified as steps per second (sps) as opposed to revolutions per minute (rpm). Eqs (6-44) and (6-45) below show the conversion from step angle to steps per revolution, and from (sps) to (rpm):

$$\text{steps per revolution} = \frac{360}{\text{step angle (degrees)}} \tag{6-44}$$

$$\text{rpm} = \frac{60(\text{sps})}{\text{steps per revolution}} = 0.167 \,(\text{sps}) \left(\frac{\text{step}}{\text{angle}}\right). \tag{6-45}$$

Step motors are available which can be run as fast as 5,000 sps (1,500 rpm for a 1.8° step motor) with a torque output of 100 inch-ounces (6.25 inch-lb). Torque output of the step motor drops off with speed for the same reason as it does in a DC permanent magnet motor. The faster the rotor and its associated magnets rotate, the more back EMF they generate in the stator coils. Back EMF limits stator coil current flow, hence magnetic field strength, hence torque output capability.

Figure 6-30 shows a typical step motor torque-speed curve. The EFSS (error-free start stop) curves show the maximum speeds and torques at which the motor may be operated without overshooting the desired final step in the absence of acceleration and deceleration speed control. The EFSS curves can be improved upon by acceleration control. Increased system inertia will also tend to degrade the EFSS curves even further to the left.

A step motor can be used in open-loop fashion to drive a positioning device (e.g., a table) as long as the EFSS torque–speed characteristic is not exceeded. If a control package (translator or indexer) employs acceleration and deceleration control, somewhat better (faster) open-loop positioning performance can be attained. Open-loop control is implemented on a machine by inputting the required number of pulses to the translator, then assuming that the step motor responded with an equal number of steps. If the step motor misses a step, or if a large inertia load causes the motor to overshoot, then actual table position is not what it should be. Each stepping error accumulates and total error will eventually affect machine performance.

Step motors are often used in closed loop drives to improve accuracy. Usually an incremental encoder is used on the same shaft as the step motor to put out a single pulse with every motor step. The first input pulse to the step motor drive results in an output pulse from the encoder. These output pulses are then used to gate all further motor pulses, i.e., no additional pulses are permitted to the motor until actual rotation as a result of the previous pulse has been acknowledged.

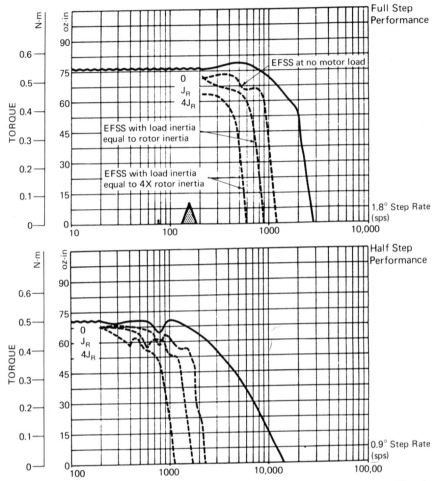

Fig. 6-30. Typical step motor torque–speed characteristic. (Courtesy of Bodine Electric Company, Chicago, IL)

6.4.5. Servomotors *(has feed back system)*

The primary use of servo motors in automatic machinery is for fast, accurate positioning of work or tools. Also important is the application of servomotors for velocity control such as that required to maintain proper feed velocity of work with respect to tool in numerically controlled machining.

Although AC servomotors are gaining importance, this discussion concentrates on DC servomotors, in particular, DC permanent magnet servomotors. DC servomotors used in machinery systems offer the following advantages over step motors:

- *Higher positional accuracy* is possible, limited only by the resolution of the position encoder. Step motors are limited by the accuracy of each step. $n_s = 120 \, f/p$
- *Higher speeds* are possible (up to 3000 rpm) with significant torque output for acceleration and deceleration.
- *Higher torque* at all speeds is available to accelerate high inertial loads. Up to 100 ft-lb of torque capability.

Permanent magnet servomotors are available with ratings ranging from fractional horsepower to 25 hp. A servomotor is characterized by a high torque-to-armature-inertia ratio for rapid acceleration and deceleration response. The designer usually tries to select a servomotor with an armature moment of inertia approximately equal to the load inertia reflected back to the motor shaft (see Fig. 6-11).

Fig. 6-31 diagrams a typical servodrive. Today, the designer can usually obtain complete servoelectronic packages to control the motor which includes power supply, amplifier, and controller. Also, DC servomotors can be purchased with integral encoder (or resolver) and/or tachometer. The primary job of the machine designer is to specify motor requirements.

The first step in specifying servomotor requirements is to calculate friction torque, viscous torque, and moment of inertia of the load reflected back to the motor shaft. Moments of inertia for typical shapes are shown in Fig. 6-10 and the reflected inertias are shown in Fig. 6-11.

The next step is to determine motor speed (rpm) requirements based on machine requirements. Assume that an *x-y* table is being driven by a DC servomotor on each axis. The required machine production rate is 2 units per second, so the table must position itself and the workpiece twice every second, plus it must dwell at each position for some amount of time, say one-quarter second per location. The distance which the table must travel between operations varies from one-half inch to six inches, with one inch or less being most common.

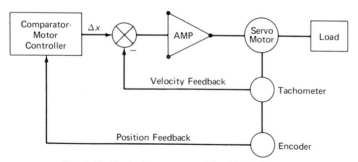

Fig. 6-31. Typical servomotor drive block diagram.

Here, the designer has to use some judgment. If he designs the table drive so that it positions the table through six inches in one-quarter second, then the motor will be oversized for normal operation requiring only one inch of travel. In order to meet the machine production rate requirement, the correct distance to use is the *average* distance travelled. Over an extended period of time, the mean production rate should then be correct.

Figure 6-32(a) shows the most practical velocity profile for the table excursion. Excess torque is applied to the machine shaft, so that the table (and motor) accelerates to maximum speed. At this point, torque drops off to the level required to match machine friction. As the table approaches its desired position, motor voltage cuts back and reverses as required to decelerate the table to zero velocity. For purposes of motor selection, table motion must be translated into motor motion at the motor shaft as shown in Fig. 6-32(b).

The velocity profile of Fig. 6-32(b) is divided into three time segments as follows:

$$t_1 = \frac{\omega_{max}}{\alpha_1} \quad \frac{(rad/sec)}{rad/sec^2} \tag{6-46}$$

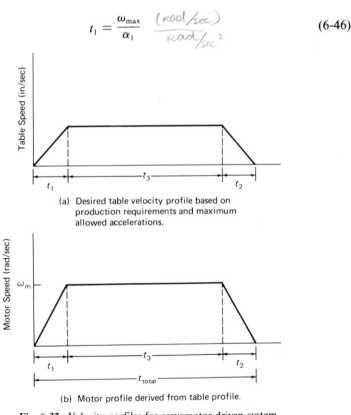

(a) Desired table velocity profile based on production requirements and maximum allowed accelerations.

(b) Motor profile derived from table profile.

Fig. 6-32. Velocity profiles for servomotor driven system.

$$t_2 = \frac{\omega_{max}}{\alpha_2} \qquad (6\text{-}47)$$

$$t_3 = t_{total} - t_1 - t_2 \qquad (6\text{-}48)$$

where: t_1 = Time to accelerate to ω_{max} (sec)
ω_{max} = Maximum motor speed (rad/sec)
α_1 = Angular acceleration of motor (rad/sec^2)
t_2 = Time to decelerate to 0 from ω_{max} (sec)
α_2 = Angular deceleration of motor (rad/sec^2)
t_3 = Time motor runs at ω_{max} (sec)
$t_{total} = t_1 + t_2 + t_3$
= Total time for table to move (sec).

Combining Eqs. (6-46) through (6-48) with the expression for the total angle θ traversed (total area under velocity–time curve):

$$\theta = \frac{\omega_{max} t_1}{2} + \omega_{max} t_3 + \frac{\omega_{max} t_2}{2} \qquad (6\text{-}49)$$

yields an expression for maximum speed:

$$\omega_{max} = \frac{2t - \sqrt{4t - 80[(1/\alpha_1) + (1/\alpha_2)]}}{2[(1/\alpha_1) + (1/\alpha_2)]^2} \qquad (6\text{-}50)$$

The next step in selecting the servomotor requires calculation of the motor torque required to accelerate the table (and motor) according to the desired velocity profile. This is found by:

$$T_{acc} = 12\alpha \, (I_{load} + I_{Motor}) \qquad (6\text{-}51)$$

where: T_{acc} = Accelerating torque required (in.-lb)
α = Maximum acceleration or deceleration from velocity profile (rad/sec^2)
I_{Load} = Moment of inertia of load reflected back to motor shaft plus inertia of drive shaft attached to motor (lb-ft-sec^2)
I_{Motor} = Moment of inertia of motor armature as specified in manufacturer's data sheet (lb-ft-sec^2).

The total torque which the motor must develop is:

$$T_{Motor} = T_{acc} + T_{fric} + T_{Load} \qquad (6\text{-}52)$$

where: T_{fric} = Friction torque at maximum speed (in.-lb)

$\quad\quad T_{Load}$ = Torque equivalent of a machine load, e.g., the positioning table lifts the weight of the table plus work against gravity (in.-lbs).

The designer now has a known point on the required torque–speed curve of the motor being selected, that is, ω_{max} [Eq. (6-50)] and T_{Motor} [Eq. (6-52)]. Recall that the torque-speed curve of a DC permanent magnet motor is approximately linear [see Eq. (6-42) and Example 6-6]. As motor terminal voltage is lowered, the motor torque–speed curve shifts inward, remaining linear and parallel to the higher voltage curve.

Manufacturer's servomotor data sheets usually show torque–speed curves at maximum terminal voltage. The designer can select a motor for which his desired ω_{max} and T_{Motor} fall on or beneath the published motor performance curve and be relatively sure that the motor will handle the job.

6.5. FLUID POWER ACTUATORS AND COMPONENTS

Even the most complex machine or workpiece motions in automatic machinery originate from actuators providing either rotary or linear motion. Section 6.4 discussed the rotary actuator most often used in industrial machine applications, the electric motor.

The most widely used linear force actuator in automatic machine design is the *cylinder*, either pneumatic or hydraulic. If one does not include the equipment necessary to produce the working fluid medium, i.e., compressed air or high pressure hydraulic fluid, then it can be said that fluid power actuators offer a much higher ratio of output power to actuator weight and volume than do electric motors.

In general, pneumatic cylinders provide faster action than hydraulic cylinders, but pneumatic cylinders are two-position devices—that is, piston and rod are fully extended or fully retracted. Although possible, it is not practical to use pneumatic cylinders as continuously variable position devices.

Hydraulic cylinders are slower in operation than pneumatic cylinders, but capable of producing very high output forces. In addition, this force can be applied smoothly over a range of ram positions, and accurately stopped at intermediate positions between fully extended and fully retracted piston. Continuous positioning of large loads is thus feasible with hydraulic cylinders.

Although cylinders are by far the most common fluid actuators, continuous rotation fluid actuators (motors) are also available. Section 6.5.4 discusses these devices. Section 6.5.5 discusses other fluid devices which find use in the design of automatic machines.

6.5.1. Pneumatic Cylinders

Pneumatic cylinders are popular actuators for use in automatic machinery applications. Most industrial facilities have shop air at 80–100 psig readily available. Low-pressure air can be exhausted directly to the plant environment after use in powering pneumatic devices without posing a safety hazard.

The fundamental application of a cylinder is to exert a linear force through a distance known as the cylinder *stroke*. A pneumatic cylinder exerts a nonuniform force over the extent of its stroke as a direct consequence of the compressibility of air.

Fig. 6-33 shows the various common cylinder types and associated nomenclature and graphic symbols. The cylinder most often used in automatic machinery is the double acting, single rod cylinder.

Selection of the proper pneumatic cylinder in a machine application requires consideration of the following criteria:

1. Force,
2. Stoke,
3. Speed,
4. Cylinder configuration,
5. Mounting,
6. Rod size.

$$P = \frac{F}{A} \times \frac{\pi}{4} D^2$$

Cylinder Force Requirements. Section 6.3 discussed the importance of defining machine requirements in terms of how much force must be exerted through what distance at what speed. Having defined this force requirement (and having determined that a pneumatic cylinder is the appropriate actuator), the minimum cylinder diameter can be selected by:

$$D_{min} = 1.128 \sqrt{\frac{F_{max}}{P_{min}}} \qquad (6\text{-}53)$$

where: D_{min} = Minimum cylinder diameter capable of providing required force (inches)

F_{max} = Maximum force required to be exerted by actuator (lbs); force must include any "stiction" which must be overcome to start the load in motion

P_{min} = Minimum pressure available to the cylinder (psig) after accounting for pressure losses in connecting lines.

Note that Eq. (6-53) applies to the case where pressure is applied to the full piston diameter, i.e., the blind end. Also note that cylinder nominal diameters are in fact *bore* diameters; thus D_{min} is the cylinder size as specified in the manufacturer's catalog.

If the cylinder size (diameter) is to be selected based on a force requirement,

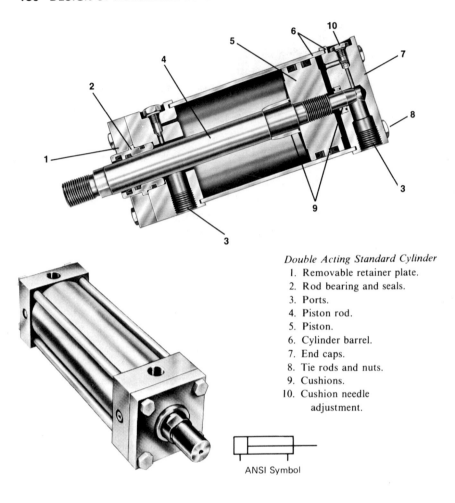

Double Acting Standard Cylinder
1. Removable retainer plate.
2. Rod bearing and seals.
3. Ports.
4. Piston rod.
5. Piston.
6. Cylinder barrel.
7. End caps.
8. Tie rods and nuts.
9. Cushions.
10. Cushion needle
     adjustment.

ANSI Symbol

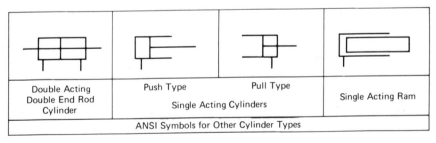

Double Acting Double End Rod Cylinder	Push Type	Pull Type	Single Acting Ram
	Single Acting Cylinders		
ANSI Symbols for Other Cylinder Types			

Fig. 6-33. Cylinder nomenclature and symbols. (Courtesy of Milwaukee Cylinder Co., Cudahy, WI)

F_{max}, such that the air pressure is applied to the *rod end* of the cylinder, then the following equation applies:

$$D_{min} = \sqrt{D_r^2 + 1.273 \frac{F_{max}}{P_{min}}} \qquad (6\text{-}54)$$

where: D_r = Diameter of the piston rod (inches).

In selecting a cylinder diameter, it is usually best to oversize rather than undersize. Too often, loads are greater than originally predicted, and line pressure is lower than normal. As a result, the cylinder either does not move the load at all, or at best, much slower than desired.

Cylinder Stroke Requirements.   Cylinder stroke is the distance between the fully extended and fully retracted rod end. Stroke is a very common catalog specification, with almost any desired stroke available.

However, in designing a cylinder drive, the selected cylinder stoke should not be much longer than that required to perform the desired machine operation. Long strokes imply long rods which are prone to (a) buckling under large loads, and (b) excessive deflection under side load which accelerates seal wear. Also, many cylinders have built-in air cushions which minimize the shock load and associated machine vibration which is common when noncushioned pistons are cycled hard against their internal stops. If the cylinder stroke is much larger than the required machine stroke (i.e., the cylinder stroke is limited by position sensors so that full piston travel is not allowed) then the built-in cushioning is not realized.

As a general rule of thumb, the next standard cylinder stroke above the machine requirement should be selected.

Cylinder Speed Requirements.   Pneumatic cylinder speed is limited by (1) air inlet and outlet restrictions, and (2) load applied to piston rod. Cylinder speed is actually adjusted by metering valves located in the air inlet line or in the air exhaust line. Cylinder speed control is referred to as *metered in* or *metered out,* accordingly. Maximum nonmetered speed is, however, limited by the cylinder inlet connection size, the supply line size and length, and the control valve size.

Since air is compressible, the cylinder speed also depends on applied load, that is, the more lightly loaded the cylinder is, the faster it will extend for a given inlet pressure. The general rules of thumb for pneumatic cylinder speed are:

1. For "average" speed, size cylinder so that there is a 25% excess force (force — pressure × piston area) over maximum load.

2. For "fast" speed, size cylinder so that there is a 100% excess force over maximum load (i.e., pressure × piston area = 2 × maximum load).

Exact prediction of pneumatic cylinder speed is very difficult because it depends on the cylinder size, applied load, and control valves used with the cylinder. Fig. 6-34 shows some typical cylinder speeds for various size cylinders using $\frac{1}{4}$ inch, $\frac{3}{8}$ inch, and $\frac{1}{2}$ inch, four-way control valves. The curves assume a 100 psig supply pressure at the control valve inlet and a cylinder load equal to 50 psi times the piston area. In other words, the cylinders are sized for "fast" speed, with 100% excess cylinder force over applied load. Maximum cylinder speed realized with no applied load is approximately 1.5 times the indicated value.

Figure 6-34 is only an approximate relationship, based on an average composite of published cylinder characteristics by several manufacturers. For specific applications, manufacturers can provide more accurate cylinder speed data based on a complete system description.

Cylinder Configuration.   Selection of the type of cylinder to be used in a specific application depends on the application. Pneumatic cylinders are almost exclusively used as two position devices in automatic machinery, operating between the two extremes of piston travel. Piston return is a necessary function, thus careful consideration should be given to the type of return used. Single acting cylinders with gravity return are unreliable due to the possibility of hang up from a sticky shaft, cocked shaft, or partially plugged outlet. Single acting cylinders with spring return are acceptable for short stroke applications, but unreliable for long strokes. Long springs can hang up, misalign, lose strength with age, or fail in fatigue.

The best approach to piston return is to use a double-acting cylinder, in which the piston is driven both ways by applying full pressure to the appropriate piston side.

Location of the cylinder on a machine within restricted space allotment is a common design problem. For very short stroke requirements, the "pancake" cylinder configuration offers a compact alternative to the normal cylindrical configuration. These cylinders often have diaphragms instead of conventional sliding pistons.

Long stroke cylinders often include a *stop tube* to reduce bearing loads on the piston rod. The stop tube simply restricts piston travel all the way to the rod end cap, increasing the axial distance between rod bearing and piston when the rod is fully extended. This results in a cylinder somewhat longer than one without a stop tube, for the same stroke. The need for a stop tube can be reduced by limiting piston travel (e.g., with appropriately positioned limit switch) in a particular design to less than full extension, thereby, creating the

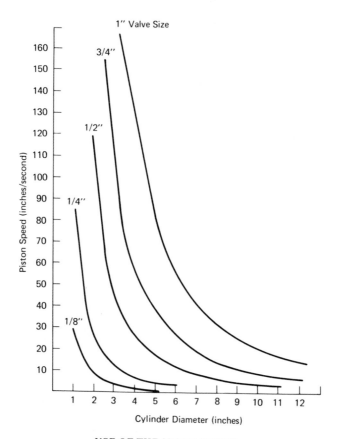

USE OF THE SPEED CURVE

1. Curves as shown are based on 100 psig supply pressure applied to a cylinder with $\frac{1}{2}$ loading, which means that the load which the cylinder is to move is $\frac{1}{2}$ of 100 psi times the piston area for the cylinder in question.

2. For cylinder loading other than $\frac{1}{2}$, apply the following factors to values of speed read directly from the graph:

0 loading   Multiply speed by 1.5
$\frac{1}{4}$ loading   Multiply speed by 1.2
$\frac{3}{4}$ loading   Multiply speed by 0.7

3. All curves assume a directional control valve in combination with the cylinder with an actual valve orifice opening equal to the valve port sizes shown ($\frac{1}{8}$, $\frac{1}{4}$, $\frac{1}{2}$, $\frac{3}{4}$, and 1 inch NPT).

4. The speed of a pneumatic cylinder depends on the type of cylinder, its particular design, and the control valve used with the cylinder, therefore the speed graph yields only approximate results. It is a composite of various manufacturer's published cylinder speed data.

Fig. 6-34. Pneumatic cylinder speed determination.

same effect as a stop tube. Also, external guides to reduce side loads on long stroke cylinders also reduce the need for stop tubes. Consult cylinder manufacturer for specific stop tube requirements for a given application.

Most cylinder manufacturers now offer cylinders with integral limit switches. A magnet on the piston actuates a reed switch at the desired cylinder extension and retraction.

The reed switches are adjustable along the cylinder length for piston stroke adjustment. These cylinders are extremely useful in situations where conventional limit switch placement is difficult due to space limitations or limited access to the moving machine member.

Cylinder Mounting. The primary cause of cylinder failure is improper cylinder mounting. The cylinder must be mounted so that the vector resultant of all load reaction forces is in line with the cylinder thrust axis (which is the theoretical center line of the cylinder bore). If the cylinder force vector and the load reaction vector are not co-linear, side forces result which rapidly wear rod seals and bearings. Note that small misalignments are magnified when the rod is in its extended position.

Cylinder mounts are either fixed or pivoted. Fixed mountings are used when the load reaction force is linear, in opposition to the cylinder thrust. Pivoted mounts (trunnion or clevis) are used when the load reaction is along a curved path, or when misalignment in the plane normal to the cylinder pivot axis is inevitable. Fig. 6-35 shows the standard NFPA (National Fluid Power Association) mounting types.

When applying fixed mounting schemes, it is best to assure centerline symmetry by using flange mounts or centerline lug mounts. End lug or side lug mounts result in combined shear and tension loads on mounting bolts, and exert a bending moment on the cylinder structure itself.

The primary design objective in mounting a cylinder is to assure that rod side loads do not occur (or are minimized) over the *entire piston travel.*

Cylinder Rod Size Selection. The object of cylinder rod size selection is to assure that the rod does not buckle under load at full extension. The criteria for buckling under compressive load depends not only on rod diameter and unsupported length, but also on the cylinder mounting and rod support conditions.

If the effective rod length-to-diameter ratio is less than 6:1, then a rule of thumb for rod diameter required is that the compressive stress in the rod be less than 20,000 lbs/in.^{2}. Effective rod length is essentially the distance between the extended rod end and the cylinder mounting point. Rod diameter considerations are not usually critical in pneumatic cylinder applications because compressive loads are low. Hydraulic cylinders under heavy load present the major problem.

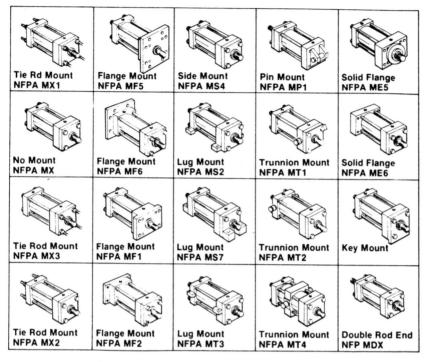

Tie Rd Mount NFPA MX1	Flange Mount NFPA MF5	Side Mount NFPA MS4	Pin Mount NFPA MP1	Solid Flange NFPA ME5
No Mount NFPA MX	Flange Mount NFPA MF6	Lug Mount NFPA MS2	Trunnion Mount NFPA MT1	Solid Flange NFPA ME6
Tie Rod Mount NFPA MX3	Flange Mount NFPA MF1	Lug Mount NFPA MS7	Trunnion Mount NFPA MT2	Key Mount
Tie Rod Mount NFPA MX2	Flange Mount NFPA MF2	Lug Mount NFPA MT3	Trunnion Mount NFPA MT4	Double Rod End NFP MDX

Fig. 6-35. NFPA cylinder mounting nomenclature.

As the effective rod length-to-diameter ratio increases above 6:1, the allowable compressive stress in the rod becomes considerably less. The rod buckling phenomenon is not a material failure, rather it is a structural instability and depends to a certain extent on the design of the specific cylinder. For this reason, it is wise to follow manufacturer's recommendations regarding rod size for long stroke, highly loaded cylinders.

6.5.2 Hydraulic Cylinders

Hydraulic cylinders are designed for use with hydraulic fluid at 2000–5000 psig as the working medium. Unlike pneumatic cylinders where exhausted fluid (air) can be simply vented to the environment, the exhaust fluid must be returned at low pressure to a hydraulic fluid reservoir. Hydraulic systems require a self-contained hydraulic pump to generate working fluid pressures and flow rates, thus are more complicated than pneumatic systems which generally use shop air directly.

On the positive side, hydraulic cylinders are capable of moving much greater loads than comparably sized pneumatic cylinders, and the associated acceleration and velocity of motion can be controlled more precisely than in the pneumatic counterpart. Hydraulic systems are generally safer than pneumatic systems even though working pressures are much greater. Hydraulic fluid, which is virtually incompressible, cannot store energy as does compressed air. A high pressure hydraulic hose rupture results in little more than spillage of fluid following an immediate pressure drop due to the rupture. A ruptured air pressure hose can propel debris for considerable distances or gyrate wildly as air escapes.

Selection of hydraulic cylinders for a particular application requires consideration of the same six criteria as in selection of pneumatic cylinders: force, stroke, speed, cylinder configuration, mounting, and rod size.

Cylinder Force Requirements. Equations (6-53) and (6-54) can be used for sizing hydraulic cylinders. Pressures will be an order of magnitude higher than those used in pneumatic systems, so forces will be proportionally higher. Care must be taken to assure that all inlet and outlet hoses and connectors are rated for the maximum level of pressure available from the hydraulic pressure unit.

Cylinder Stroke Requirements. Hydraulic cylinders are often used for very long stroke applications as in machine tool table feeds, or in material transfer elevators. Side loading is the major contributor to hydraulic cylinder failure, as it is in pneumatic cylinders. Therefore, stop tubes should be used or cylinders should be sized longer than the design stroke, and the rod travel restricted by controls to less than full extension.

Cylinder Speed Requirements. Hydraulic cylinder speeds are extremely controllable and predictable, unlike pneumatic cylinder speed. Hydraulic pumps are positive displacement devices, putting out a constant *flow rate Q* rather than a constant pressure. The hydraulic pressure depends on the degree of flow restriction downstream of the flow source (pump). Hydraulic cylinder piston speed is limited by the flow rate of hydraulic fluid into the cylinder as follows:

$$V_e = 1.273 \, Q/D^2 \qquad (6\text{-}55)$$

where: V_e = Cylinder piston speed (in./sec) for rod extension.
Q = Flow of fluid into cylinder (in.^{3}/sec)
D = Hydraulic cylinder bore diameter (in.)

Hydraulic cylinders are generally slower than pneumatic cylinders because the working fluid must be pumped into the cylinder in order to move it. In pneumatic cylinders, the air expands into the cylinders. If the air were not restricted, the theoretical speed of pressure wave expansion would be the speed of sound (in air at operating temperature).

Because hydraulic pumps output a constant flow rate through a given load, the hydraulic piston extends smoothly, at constant velocity. Speed can be controlled by varying fluid flow rate to the cylinder by means of flow control valves. Many cylinders have built-in flow controls to adjust cylinder speed.

In a double-acting single-rod-end cylinder, the reverse stroke is always *faster* than the forward stroke for the same hydraulic fluid inlet flow rate. Because the volume behind the rod end of the piston is reduced by the volume of the rod itself, this "cavity" must expand faster (i.e., retract the rod) for the same flow rate into the cylinder. The hydraulic cylinder retraction speed is thus:

$$A = \frac{\pi}{4}\left(D^2 - D_R\right)$$

$$V_r = 1.273Q/(D^2 - D_r^2) \tag{6-56}$$

where: V_r = Cylinder piston speed (in./sec) for rod retraction, single rod end cylinder (note that V_r is also the piston rod extension and retraction speed for a *double*-rod-end cylinder)

D_r = Piston rod diameter (in.).

Cylinder Configuration.   Selection of hydraulic cylinder configuration involves the same considerations given pneumatic cylinders, and is usually a function of the specific design requirement. Most applications use double-acting, single-rod-end cylinders. One notable exception is the *hydraulic ram,* where piston rod diameter is the same (or slightly smaller) than the piston itself. Rams are used to apply very large forces, as in hydraulic presses or lifts. Piston return is effected by the dead weight load on the ram (e.g., in a lift) or by an opposing ram.

In general, spring piston return is not used in hydraulic cylinders because the spring forces are small in comparison to the hydraulic pressure forces. Also, hydraulic cylinder applications usually require the load to be driven in both directions, so that double-acting cylinders are appropriate.

Double-rod-end cylinders offer the design equivalent of two cylinders in one, provided the requirement is for colinear motion. Forward and reverse strokes are at the same speed because the effective areas of both pistons are reduced by the rod area. Double-rod-end cylinders have been used to solve difficult limit-switch placement problems because the limit-switch-actuating cam can be placed on the nonworking rod, out of the way of the working rod.

Cylinder Mounting.   Care in the mounting of hydraulic cylinders is essential because of the high loads usually associated with hydraulic applications. Misalignment of the cylinder mounting axis with respect to the load application axis results in side loads on the piston rod which vary with piston position. Side forces may increase or decrease with rod extension, depending upon the initial misalignment. The rod bearing and seal is better able to support side loads when the cylinder is retracted because this condition results in a wide support base between rod bearing and piston bearing (in the cylinder bore). However, side loading due to misalignment should never be tolerated. Cases where side loads are inevitable require a rethinking of the entire design, or acceptance of periodic cylinder replacement as a required machine maintenance activity.

Cylinder Rod Size Selection.   Rod size selection is important because hydraulic cylinder stokes and loads tend to be higher than in their pneumatic counterparts. The rod diameter must be large enough to prevent buckling at the fully extended position under the maximum possible load. The maximum possible load will be experienced if the rod and its associated machine member is stopped (e.g., a jam-up against a solid stop). The load transmitted through the rod will then be the maximum fluid pressure times the piston area.

Buckling is a structural instability, and depends on the cylinder mounting scheme and cylinder design in addition to the rod size. It is thus important to follow manufacturer's recommendations for rod size selection, since their data is based on actual tests with the hardware being considered.

6.5.3. Cylinder Directional Control Valves

Cylinders, hydraulic or pneumatic, are never found in automatic machine applications without an associated *directional control valve*. The directional control valve provides a means (either manual or automatic) for applying the pressurized working fluid to the cylinder, and for exhausting low pressure fluid once cylinder work has been done. Since cylinders are never used without the associated directional control valve, such valves will be treated in this section as an integral part of the cylinder.

Directional control valves provide a means of directing the high pressure at one or two valve inlets to one, two, or three valve outlets in response to a *control actuation*. The control actuation can be manual, electrical solenoid actuation, mechanical cam actuation, or pressure actuation. Depending on which valve outlet the high pressure is directed to, the associated cylinder extends, retracts, or stops (in mid-stroke). Clearly, the cylinder would not be a useful automatic machine component if it could not be controlled.

Directional control valves divert working fluid to appropriate outlet ports by sliding spools, plungers, poppets and diaphragms, to mention the common types. Each valve type has its advantages for specific applications, but the *spool-type* directional valve has a very wide range of applicability and will be used exclusively in the discussions which follow. Conventional fluid power symbology is based on the sliding spool-type valve, and fluid circuit operation is easily visualized with the spool valve concept in mind (even though the actual valve technology used might be of a different type).

Figure 6-36 shows a two-position, four-way shuttle or spool valve. The

PRINCIPLES OF OPERATION

DE-ENERGIZED

When the V5 type solenoid operator is in the deenergized position, the pilot inlet orifice, 1, is sealed by the soft insert 2, in the bottom of plunger 3. The spool 4, is so designed that the pressure port, P is open to the B cyl port and the A cylinder port is open to the exhaust port, EA.

As a result, pressure is supplied to B cylinder while A cylinder is exhausting through port EA. Port EB is sealed off by the spool and "O" rings.

The spool in the valve body is held in this position by means of a mechanical spring and air pressure in the spring cavity. The air pressure is supplied through a passageway, 6 that intersects the pressure port and connects to the pilot orifice 1, to provide pilot pressure.

The "O" rings, 7 prevent interport leakage and the end "O" rings, 8, prevent external leakage.

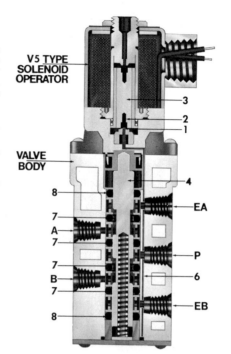

ENERGIZED

When the solenoid coil, 9 receives current the resulting magnetic flux pulls the plunger 3, up against the stop 10 opening pilot orifice, 1. The insert, 11 in the top of the plunger seals the pilot exhaust orifice. The pilot pressure provided through passageway, 6 flows through the pilot orifice and through a passageway not shown to the back of the motor assembly. The force generated by the pilot pressure (at least 15 PSI) against the motor assembly piston is sufficient to drive the spool, 4 downward, compress the machanical spring 5, and overcome the air pressure in the mechanical spring cavity and open pressure to A cylinder and B cylinder to exhaust port EB. The spool and "O" rings will seal port EA.

When the valve is de-energeized, the pilot plunger and inserts will close off the pilot pressure and open the pilot exhaust orifice—the mechanical spring and air pressure will return the spool to its original position. The pressure in back of the motor assembly will be exhausted out the pilot exhaust orifice located in the sleeve.

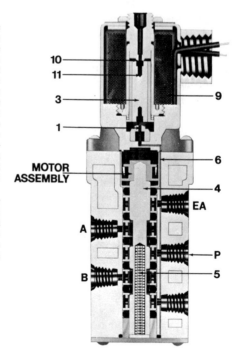

Fig. 6-36. Two-position, four-way spool valve. (*Courtesy Skinner Precision Industries, Inc., New Britain, CT.*)

spool in a two-position valve can assume only two positions—fully left or fully right. When the spool is pushed to the left by the valve *actuator* or *operator,* high-pressure fluid flows from inlet port 1 to outlet port 3, and port 2 is simultaneously connected to port 4.

Figure 6-37 develops the valve symbology in common usage. Valves are diagrammed as two or three boxes, where the number of boxes represents the number of valve positions. [Fig. 6-37(a)]. The external lines leading to the valve diagram represent the fixed valve ports through which fluid may flow, and the number of such lines is the number of "ways" associated with the valve [Fig. 6-37(b)]. Two-, three-, four-, and five-way directional control valves are the standard types available.

The way in which the valve ports are connected internally through the valve are shown by lines inside the valve box diagram, with arrowheads to designate the direction of fluid flow (always from high to low pressure).

The boxes represent the valve shuttle or spool, and are movable to as many positions as there are individual boxes. The external lines, on the other hand, represent fixed ports in the valve body, and indeed, represent actual fluid connections from the valve to other fluid power elements. Fig. 6-37(c) shows the manner in which external ports (lines or "ways") are interconnected for each of the two valve positions.

The position of the spool with the valve body is determined by the valve actuators. The actuators are shown symbolically as being attached to both ends of the valve spool. Regardless of how the valve actuator actually works, it is always interpreted in the diagram as pushing the spool in order to line up internal and external fluid passages. Fig. 6-38 denotes the most common valve and actuator symbols.

Assume a double-acting cylinder used in conjunction with a single-solenoid two-position, four-way valve. In its unenergized state, the valve supplies air to the rod end of the cylinder. When the electric solenoid is energized, it pushes the spool to a position such that high pressure air flows to the blind end of the cylinder, causing it to extend. When the power to the solenoid is removed, the spring returns the valve to its non-energized position, and high pressure air flows to retract the cylinder to its home position. Note that the cylinder piston will always start out from its retracted position (i.e., before the solenoid is energized) and in the event of power failure, the cylinder would automatically return to its home position.

The following rules summarize the drawing and use of directional valve diagrams in fluid power circuits:

1. All external connections from the valve to the rest of the fluid circuit must be made through only one box of the cylinder diagram.
2. All connections to the external fluid circuit from the valve diagram should be shown for the valve in its non-energized or non-actuated state.

3. Visualize all valve operators as pushing the spool diagram. Therefore, show all external connections to the box farthest away from the operator. (When the operator is actuated, it will push the spool so the box nearest the operator lines up with external connections.)
4. Arrowheads inside the boxes show the direction of flow through the valve when that box is in its working position, i.e., connected to the external circuit.

Directional control valves are diagrammed as two or three boxes, where the number of "boxes" represents the number of valve positions. The series of two or three boxes are thought of as the movable part of the valve, analogous to the spool in a spool-type valve:

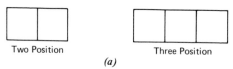

Two Position   Three Position

(a)

Fluid lines external to the valve are diagrammed as lines connecting to one of the boxes of the valve diagram. The number of such lines define the number of "ways" fluid flows into and out of the valve, thus define the number of ways of the valve:

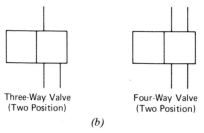

Three-Way Valve   Four-Way Valve
(Two Position)   (Two Position)

(b)

Arrows drawn inside each of the valve boxes show the direction of fluid flow through the valve (always from high pressure to low pressure) when the valve spool is in the position which lines that box up with the external lines:

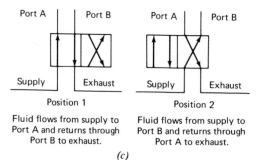

Port A   Port B   Port A   Port B

Supply   Exhaust   Supply   Exhaust

Position 1   Position 2

Fluid flows from supply to   Fluid flows from supply to
Port A and returns through   Port B and returns through
Port B to exhaust.   Port A to exhaust.

(c)

The method by which the valve spool is moved from one position to another is diagrammed by a symbolic actuator (see Fig. 6-38) placed at one or both ends of the valve diagram (i.e., array of boxes). The actuator is thought of as pushing the valve when the actuator is energized. The

Fig. 6-37. Steps in diagramming fluid power directional control valves.

external lines are always shown attached to the box to which they would be connected ɪɪ the actuator were unenergized. Therefore the actuator is always diagrammed on the end of the valve *away* from the box to which the lines are attached. (*Exception:* When there are two actuators, one on each end of the valve, the unenergized valve position is that which the valve would normally be in when the fluid power circuit is turned OFF).

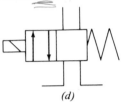

(d)

Two-position, four-way, single-solenoid-actuated valve diagram, Note that lines are shown connected to box away from solenoid actuator. This is the position that the valve would be in if the solenoid was OFF. The spring would push the valve into the position shown in the diagram.

Fig. 6-37. (*Continued*)

Application of fluid cylinder and directional valve combinations in automatic control systems is demonstrated in Chapter 9.

6.5.4. Fluid Motors

Fluid motors differ from cylinders in that the application of high pressure fluid results in continuous rotary motion of the output shaft, much like electric motors. Air motors are used in high speed (up to 6000 rpm), low power (up to 25 hp) applications relative to hydraulic motors which typically operate at less than 3000 rpm, but which can output several hundred horsepower. Both pneumatic and hydraulic motors are used as alternative prime movers to electric motors in applications where explosion hazard due to electrical arcing may exist. Hydraulic motors are used instead of electric motors in very high load applications where large electric motor size and weight presents a design problem.

Air Motors. Air motors used in automatic machinery are typically vane type. Their primary industrial use is in driving conveyors, mixers, or blowers in explosive atmospheres. Torque versus speed characteristics are not unlike that for a series-wound DC motor, in that maximum torque is realized at zero speed (stall torque), and drops off approximately linearly with motor speed.

Air motors will "run away" if the shaft load is removed, and may, depending on their design, damage themselves. A shaft speed governor may therefore be necessary. Air motors can be noisy also, since the low pressure air is generally vented directly to the environment. Even when exhaust air mufflers are used, the noise level is higher than that of a comparatively sized electric motor.

Air motor direction of rotation can be reversed relatively easily in

Directional Control Valve	Symbol	Directional Valve Actuators	Symbol
Four-way, two-position valve		Solenoid	
Three-way, two-position valve		Solenoid, spring centered	
Two-way, normally closed valve		General purpose	
Two-way, normally open valve		Pilot operated	
Typical four-way, three-position valve		Button bleed	
Three-position valve center porting, load isolated, power shunt		Manual lever	
Load and power shunt		Foot pedal	
Load and power isolation		Palm button	
Hi-pressure load equalization		Cam operated	
Low-pressure load equalization		Detented	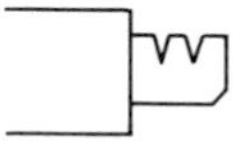
		Spring return	

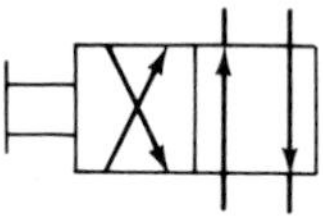

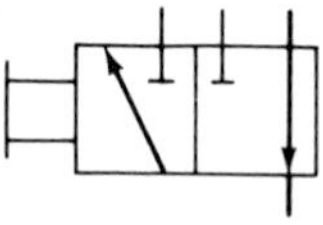

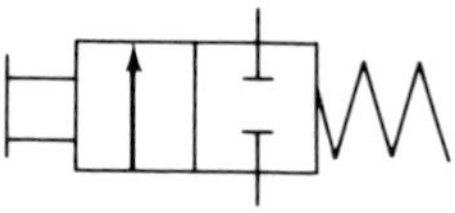

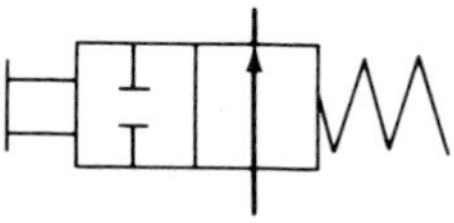

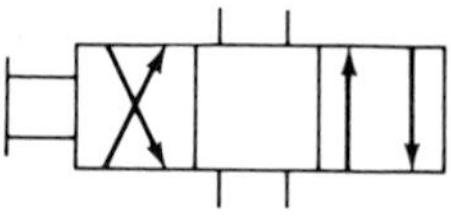

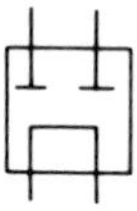

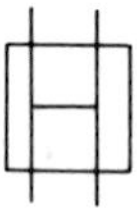

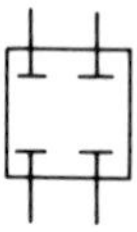

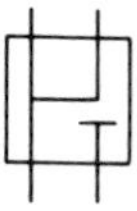

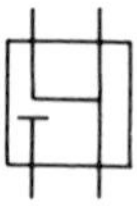

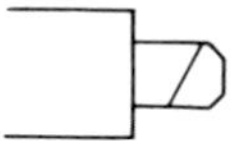

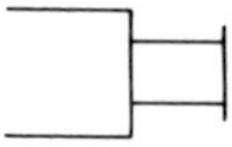

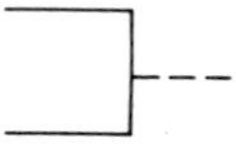

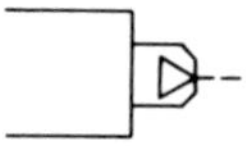

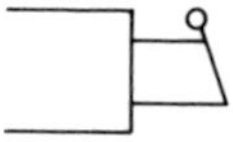

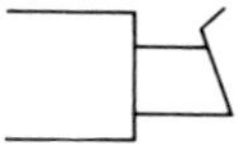

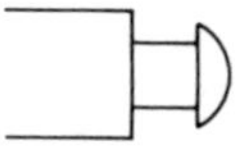

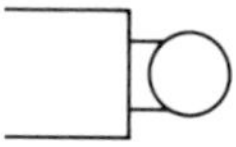

Fig. 6-38. Standard symbols for directional control valves.

conjunction with a four-way valve. Direction of air flow through the motor is reversed in the same manner as cylinder direction is reversed. Speed control of air motors however, is very difficult due to the compressibility of air and the load sensitivity of the motor. Hydraulic motors should be used if precise speed control is required.

Hydraulic Motors.   The three general types of hydraulic motor are the gear type (most common), vane type, and piston type (least common). Gear-type hydraulic motors are known as *fixed displacement* motors because motor speed is regulated by varying the flow rate to the motor. Torque output at a given inlet pressure is constant, regardless of speed. Vane-type motors are generally fixed displacement but can be designed as *variable displacement* motors where both speed and torque vary with flow rate.

Gear type hydraulic motors provide constant torque output (torque output varies with pressure drop across motor) for speeds in the range of 100–3000 rpm. Efficient low speed operation is difficult because volumetric efficiency is low due to proportionately high fluid-leaking around the gears. Vane type hydraulic motors have better low speed performance, and can be used down to 10 rpm. Both types are generally reversible using directional control valves. Speed is varied by means of flow control valves.

6.5.5. Other Fluid Devices Used in Automatic Machinery

Cylinders and fluid motors represent virtually all of the fluid power actuation devices used in automatic machinery. However, other pneumatic and hydraulic components find widespread usage.

Rotary Actuators.   Rotary actuators differ from fluid motors in that they provide intermittent rather than continuous rotation. Rotary actuators are usually constructed with one or two cylinders which provide torque output through a fixed angle (up to 720°) as the cylinder(s) extend. The load may be driven both forward and backward if a rack and pinion mechanism is used (Fig. 6-39). The load may also be intermittently driven in one direction if a ratchet and pawl mechanism or a one-way clutch mechanism is used. Hydraulic actuators are used for very high loads, and pneumatic cylinders may be used for smaller torque requirements.

Rotary tables are commonly used in the material transfer subsystem of an automatic machine and also use cylinder actuation to index the table. The purpose of a rotary table is to provide accurate angular positioning with each index rather than providing a constant torque through an angle. For this reason, rotary tables may also employ a positioning key which locks the table at an accurate location after each index.

1   Service - Air, Hydraulic, Air/Oil-Tandem.

2   Design Principle - **Rack and pinion** with **external angle adjustments** allowing infinite choice of rotation between 0 and 180 deg. **Zero backlash** at both end positions.

3   End Caps - Steel; zinc plated to prevent corrosion.

4   Pistons - Aluminum pistons are floating to enhance low friction and high efficiency.

5   Piston Seals - Liptype, pressure energized, and wear compensating for low friction and long life. Each piston has double seals to assure leak free performance under normal conditions. (See pg. 4).

6   Pinion - Steel - One piece gear and output shaft.

7   Actuator Body - Drawn Aluminum alloy; all surfaces including bearing areas are hard anodized for maximum wear resistance.

8   Port Control® - Built in flow control valves are standard on all models eliminating need for external flow controls.

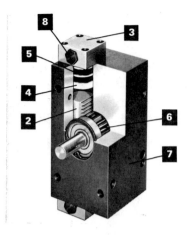

Fig. 6-39. Cylinder-powered rotary actuator rack-and-pinion drive. (*Courtesy PHD, Inc., Fort Wayne, IN.*)

Fluid Shock Absorbers. Since most automatic machinery is of the indexing type, frequent starts and stops are common. Machine accuracy requirements usually dictate that a workpiece or part must not only be stopped, but stopped at a precise location within the machine.

Accurate work or part positioning is most easily accomplished by bringing the part up against a precisely located hard stop. However, it is not good design practice to stop a part by running it or its holder up against a stop at high speed. Damage may result to the part, to the stop, or to the machine (e.g., through vibration).

One alternative to part or work speed control is to control the acceleration and deceleration of the work moving actuator (motor or cylinder). This usually requires complex control circuitry or programming however. A good way to control part deceleration, is to use a *shock absorber* or *dashpot* or *damper* to cushion its impact against a hard stop.

Most dampers are pneumatic or hydraulic, with the rate of deceleration controlled by forcing fluid (air or hydraulic) through an orifice. The kinetic energy of the moving part is thus dissipated as heat as the fluid flows (reversibly) through a restriction. A mechanical spring usually resets the damper to its extended position.

Air Jets. Air jets are used in automatic machinery to clear swarth (chips), to eject parts from nests, to assist parts motion as they are transferred in tables, to reposition parts, and to select parts (as wipers in vibratory bowl feeders), among other applications.

The air jet may be continuous or intermittent, controlled by a solenoid operated two-way (ON or OFF) valve.

Vacuum Sources.   Vacuum is used extensively in automatic machinery. The two major applications are vacuum chucking (work holding) and vacuum pick-up (for work or part transfer). It is possible to use a piston-type or diaphragm vacuum pump as a vacuum source, and indeed, this is the only alternative for very-high-vacuum requirements. The vacuum pump itself is a complicated device, however, so the method shown in Fig. 6-40 is used whenever possible.

High-pressure (line) air is allowed to flow through a venturi tube, usually by actuating a two-way solenoid valve. The air flow velocity results in a negative static pressure within the venturi throat, and this is the source of vacuum used in machinery. Note that ON–OFF control of the vacuum is relatively easy, accomplished by energizing or de-energizing a valve solenoid.

Use of any vacuum source requires evacuation of existing air from a cavity between the workpiece and the vacuum line. Since evacuation takes time, proper design requires minimizing vacuum chuck or pick-up head cavity volume, as well as minimizing vacuum line diameter. Also, leaks between the vacuum cavity and ambient air pressure must be minimized. This is accomplished by using flexible shrouds or boots on the pickup head, and by only attempting vacuum chucking or pickup of smooth, continuous surface work.

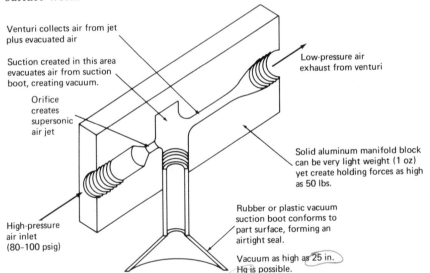

Venturi collects air from jet plus evacuated air

Suction created in this area evacuates air from suction boot, creating vacuum.

Orifice creates supersonic air jet

Low-pressure air exhaust from venturi

Solid aluminum manifold block can be very light weight (1 oz) yet create holding forces as high as 50 lbs.

High-pressure air inlet (80–100 psig)

Rubber or plastic vacuum suction boot conforms to part surface, forming an airtight seal.

Vacuum as high as 25 in. Hg is possible.

Fig. 6-40. Standard technique for generating machine vacuum.

6.6. SOLENOID ACTUATORS

Solenoids are widely used as integral actuators in other machinery or machine control elements, including directional fluid control valves, motor starters, relays, stepping switches, clutch and brake actuators, and safety shutoff valves and switches. Solenoids may also be used as separate machine components, capable of producing relatively low-force (0–100 lb) and short-stroke (0–1 inch) motion. Rotary solenoids are also available which produce torque through discrete angles of up to 100°. Machine uses of solenoids include: locking mechanisms, gating mechanisms, escapements, part ejection, and low-force positioning and/or clamping.

6.6.1. Solenoid Principles

A solenoid is a two-position electromechanical device. Figure 6-41 shows a typical linear solenoid consisting of a fixed, wirewound coil which surrounds a movable, magnetically soft *armature* or *plunger*. The coil is often enclosed externally with a magnetically soft housing, so as to concentrate the magnetic field of the coil at both ends and through the center of the coil.

The armature provides a magnetic "short circuit" through its own mass, so that virtually all of the coil field is concentrated in the air gaps at the ends of the solenoid. When the solenoid coil is energized, the armature will seek an equilibrium position within the coil which minimizes the *reluctance* of the magnetic field in the end air gaps. The tendency of the armature to seek this equilibrium is manifested by a force which tends to pull the armature into the coil. This force is greatest at the equilibrium point, and drops off approximately as the square of the distance the armature is moved out of the energized coil (i.e., away from equilibrium).

In order to make the solenoid a useful device, the armature must be offset from its equilibrium position in a de-energized coil by a spring or by gravity, so that when the coil is energized, the armature moves or "pulls in" to the equilibrium or seated position. The distance which the armature travels is known as the *stroke* of the solenoid. Note that there are only two positions possible: the initial, de-energized position and the energized equilibrium or seated position.

Fig. 6-42(a) shows the typical force versus stroke (armature position) characteristic for a solenoid. The maximum force is realized when the coil is energized and the armature is fully pulled in. This is referred to as the *holding force* of the solenoid. The minimum force is exerted immediately after coil energization when the armature is at the full extent of its stroke (i.e., initial position). Fig. 6-42(b) shows the typical solenoid pull-in time as

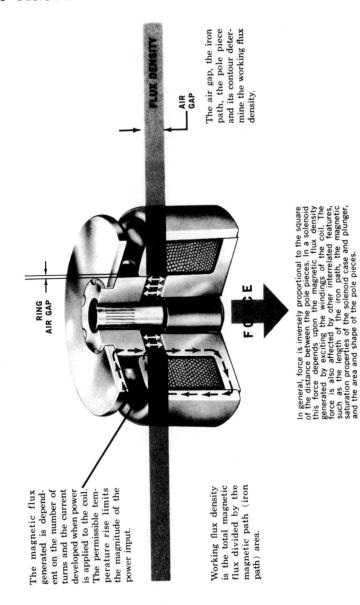

Fig. 6-41. Linear solenoid. (Courtesy of Ledex, Inc., Vandalia, OH)

FLUX DENSITY

AIR GAP

The air gap, the iron path, the pole piece and its contour determine the working flux density.

RING AIR GAP

FORCE

In general, force is inversely proportional to the square of the distance between the pole pieces. In a solenoid this force depends upon the magnetic flux density generated by exciting the windings of the coil. The force is also affected by other interrelated features, such as the length of the iron path, the magnetic saturation properties of the solenoid case and plunger, and the area and shape of the pole pieces.

The magnetic flux generated is dependent on the number of turns and the current developed when power is applied to the coil. The permissible temperature rise limits the magnitude of the power input.

Working flux density is the total magnetic flux divided by the magnetic path (iron path) area.

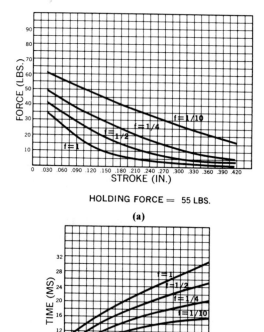

Fig. 6-42. (a) Typical linear solenoid force versus stroke for various duty cycles *f*. (b) Typical linear solenoid pull-in time versus stroke for various duty cycles. Data for same solenoid as shown in Fig. 6-42(a), no load, and 20°C coil temperature. (Courtesy of Ledex, Inc., Vandalia, OH)

a function of where in the stroke the unloaded armature starts to travel. As the armature is loaded, the time required to pull in increases. If a particular design application requires more force at startup than a particular solenoid offers, it is possible to set the initial position at something less than full stroke so that startup force is higher. However, actual working stroke is sacrificed. The distance the solenoid travels before engaging the load is known as *pretravel.*

In order for the solenoid to remain operational, i.e., pulled in, current must flow through the solenoid coil windings. The coil characteristic which generates the magnetic field, thus force, is known as magnetomotive force (or MMF) and is the number of *amp turns* of the coil:

$$MMF = NI \text{ (amp turns)} \tag{6-57}$$

where: MMF = Coil magnetomotive force
N = Number of windings in the solenoid coil
I = Current in amps passing through the coil winding.

As long as current flows through the winding, heat is generated as I^2R losses. As the coil temperature rises, winding resistance increases, thus current flow (and MMF) for a given applied voltage decreases. Holding force for a given solenoid is therefore lower if the solenoid is on continuously, than it is if that solenoid is cycled on and off intermittently. Solenoid off time allows the coil to cool down, thus a lower average temperature is realized. Solenoid force curves are usually given for several different on-off combinations or *duty cycles.*

$$f = \text{Duty Cycle} = \frac{\text{Solenoid ON Time}}{\text{ON Time} + \text{OFF Time}}. \qquad (6\text{-}58)$$

Manufacturers specify force as a function of duty cycle and also as a function of the longest ON period of that duty cycle. Obviously a coil with duty cycle of $\frac{1}{2}$ with a maximum ON time of 1 minute operates at a lower temperature than does a coil with duty cycle of $\frac{1}{2}$ and maximum ON time of one hour.

As the armature moves from its initial position to its seated position, the inductance of the solenoid coil increases. For an AC excited coil, the increase in induction results in a decrease in current flow from the *inrush* current to *rated current* or *holding current* at the seated position. This fact is particularly important in application of AC solenoids in that care must be taken to design for the armature to seat fully, and quickly. Continuous high AC current which would flow if the armature stalled at an intermediate position would result in rapid coil heating, thus loss of holding force.

6.6.2. Solenoid Selection

Solenoid manufacturers offer a wide range of stock coils which exhibit many different combinations of force/stroke characteristics, coil and frame sizes, and associated duty cycles. Although solenoids can be custom tailored for specific applications, it is always more economical to use stock coils whenever possible.

The first step in selecting a solenoid as an actuator in a machine is to define the three basic machine requirements: force, stroke, and speed. Since force and stroke are integrally related for a given coil, they must be considered together. Do not overlook the possibility of using a particular solenoid for a stroke requirement which is less than the maximum for that device. In this manner, a higher starting force will be available, although the solenoid will have some pretravel.

Manufacturers' curves usually show force versus stroke for a particular coil used at *several different duty cycles*. Light-duty-cycle application will result in higher force/stroke performance, while use of that solenoid in a continuous holding mode results in the lowest possible force/stroke performance.

Like motors, the primary cause of solenoid failure is due to coil overheating. Therefore, once a particular solenoid is selected for its force/stroke requirements, consideration must be made of coil resistance, holding current, duty cycle, maximum ON time, and the impact of each of these factors on coil temperature. Force/stroke data are usually provided by the manufacturer at one of two conditions:

1. Force/stroke curves with coil at ambient temperature (20°C);
2. Force/stroke curves with coil at maximum allowed temperature of 100°C (20°C + maximum 80°C rise due to coil heating).

If curves are provided for coils at ambient temperature of 20°C, then they may be *derated* or scaled down to allow for temperature rise by dividing the solenoid force by the factor F shown in Fig. 6-43 as a function of coil temperature. If the solenoid is to be used for continuous holding with rated voltage applied to the coil, then the maximum temperature rise of 80°C should be assumed, and the coil holding force derated accordingly.

If the ambient temperature in which the solenoid is to operate is *not* 20°C, i.e., colder or warmer, then the maximum coil operating temperature may be calculated by adding 80°C to the ambient temperature. The 80°C maximum temperature rise figure is derived from NEMA standards for Class A insulation.

If more or less than rated coil voltage is applied to a solenoid, then proportionally more or less current will flow through the coil winding. Higher voltages will give more holding power, but the coil will reach a higher operating temperature. Similarly, lower than rated voltages will run cooler, but will have less holding force.

The designer is often confronted with the dilemma of having to provide continuous holding force, but also requiring a relatively high startup or *pull-in* force. In order to obtain required pull-in force, a large solenoid must be used which provides much more holding force than necessary, and wastes power accordingly. Several techniques are employed so that a smaller solenoid may be used for the above requirement.

The design concept is to provide a pulse of high voltage to the cold solenoid coil just long enough for the solenoid to pull in. That voltage is then reduced to rated voltage or lower to *hold* the solenoid. The lower holding voltage will be sufficient to provide required holding force continuously, without overheating the coil.

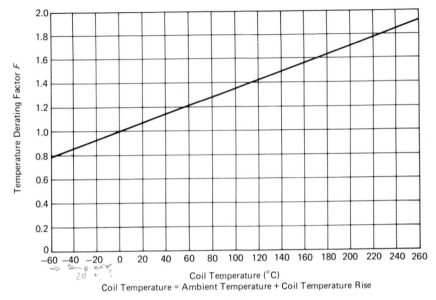

Fig. 6-43. Solenoid derating curve as a function of maximum coil temperature.

One technique is shown in Fig. 6-44(a) where a normally closed (NC) switch is placed in parallel with a voltage reducing resistor ahead of the solenoid coil. When the solenoid circuit is energized, the NC switch assures full line voltage to the coil. The pull-in motion of the solenoid armature must then mechanically open the NC switch (e.g., a limit switch) so that line voltage is reduced to the desired holding valve through the resistor. Note that this technique does not save power, but only redistributes its dissipation between the coil and resistor. However, the coil will run cooler at the reduced voltage.

Figure 6-44(b) shows a technique for applying high voltage for a time interval (presumed larger than the pull-in time) before reducing it to the holding level. Initially full current flows through the solenoid coil and the transistor base–collector circuit. As the capacitor charges, less current flows through the coil. When the capacitor is fully charged, the transistor *shuts off* and the only current path is through the solenoid coil and the *hold-in resistor*. As in the previous example, reduced voltage is applied to the solenoid coil by virtue of the resistor. Obviously, this technique requires that the capacitor be sized so that the transistor does not shut off before the solenoid armature is pulled in.

A third technique would be to apply appropriate coil voltages in response to an external control circuit. Again, timing is important so that coil voltage is not reduced prematurely.

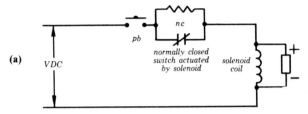

Mechanical hold-in resistor circuit.

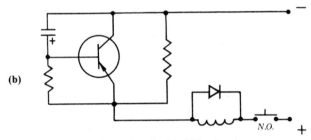

Transistorized hold-in circuit.

Fig. 6-44. (a) Mechanical method of providing pull-in pulse for a solenoid; **(b)** electronic method of providing pull-in pulse for a solenoid.

The final consideration is solenoid speed. The pull-in time for a solenoid depends on the external load on the armature, and the level of coil energization. Higher external loads will be moved slower by the solenoid, and indeed, excess heating will occur in AC solenoids if the load restricts pull-in to any extent. The pulsed technique of applying high voltage during pull-in, then reducing it for holding will speed up solenoid pull-in time considerably.

Many manufacturers provide pull-in time curves [Fig. 6-42(b)] which show pull-in time as a function of duty cycle and stroke. Pull in time versus stroke is roughly inversely proportional to force versus stroke for the same solenoid. Typical solenoid pull in times range from 5–100 milliseconds, with 10–20 ms being common.

REFERENCES

Jaeschke, Ralph (Ed.); *CONTROLLING POWER TRANSMISSION SYSTEMS,* 1978, Penton/IPC Inc., Cleveland, OH.

Metcut Research Associates, Inc., *MACHINING DATA HANDBOOK,* 3rd Edition 1980, Cincinnati, OH.

7. PRACTICAL WORK STATIONS

7.1. SELECTING OFF-THE-SHELF WORK STATIONS

Almost all production machine designs employ the *modular design* concept. The designer selects, where available, a work station actuator capable of performing the required operation on the workpiece, and integrates it into the machine by mounting it on a chassis in such a way that the workpiece can be efficiently moved into and out of position for the operation to take place.

Table 7-1 lists but a few of the standard work stations available as packages, ready to mount in a special machine. The proper design procedure is to first decide exactly what operation is required at a given work station. All potential package work station candidates should then be considered. If the designer is not sure of available packages, or if he has not used them in some time, an exhaustive search of available items should be made. One source not to be overlooked is the manufacturer of similar standard machines. Although they do not advertise machine parts, one can often purchase the critical actuator (say a grinding head) needed for a special machine without purchasing the entire standard machine.

Many times the best source of standard automated tooling are manufacturers of items associated with that tooling, e.g., rivet manufacturers for automatic rivet setting equipment, hot melt adhesive manufacturers for adhesive application heads, retaining ring manufacturers for automated snap ring application tools, etc.

The advantage of using packaged subsystems, or even other manufacturers' spare parts, is that the work station as a whole has been fully engineered, probably with many revisions, and has been field proven and fully debugged. If the designer can be sure that the group of packaged work stations which he has assembled into a complete special purpose machine will all perform the required function to the required level of quality, his only remaining task is to design the method for inserting, positioning, and removing the workpiece with respect to the work station (i.e., design the material transfer subsystem).

Often the entire special machine is built around an exotic work station,

TABLE 7-1. Off-the-Shelf Work Stations.

Type of Automatic Workhead	Operation(s) Performed	General Description
Air press	Metal bending, piercing, forming notching, cutoff staking, swaging, heat sealing	To 20 ton, 5 inch strokes, 100 psig air actuated mechanical toggle multiplies force at the expense of stroke.
Hydraulic press	Metal bending, forming, etc.	To 200 ton, very long strokes.
Drilling head	Drilling, peck drilling, tapping, milling	Electric motor (to 15 Hp) spindle, hydraulic feed
Automatic screwdriver	Feed and drive screws	Pneumatic screw feeding, torque control, wide variety of screws, rates to one per second.
Torquers	Feed and torque bolts, run nuts	Air or motor driven, torque control
Riveting head	Feed and set rivets	Electric motor, hydraulic, or air powered. Rates to 2.5 per second.
Welding heads	Welding	Resistance welds, automatic electrode feed and energize.
Soldering heads	Soldering	Automatic head feed and energize.
Ultrasonic welding	Fasten plastic parts	Use ultrasound to melt plastic and weld along narrow seams.
Hot-melt adhesive applicator	Apply hot melt adhesive	Melt and meter precise amounts of adhesive.
Automatic fill heads	Meter and fill containers with liquids	Plunger actuated, positive displacement filling.
Pin insertion	Feed and insert small pins	Pneumatic insertion, vibratory bowl feed of pins.
Automatic bagger	Feed and count parts, bag, heat seal, label	Plastic bags, rates to one per second.

such as a laser machining system (laser scriber), ion milling system, etc. In this case, the machine design job is essentially one of designing efficient material (work) transfer, positioning, clamping, and control. Note, however, that performance features such as cycle rate, quality, etc. are dictated by the work station, and are not variables under the control of the machine designer.

One of the most important specifications of the work station, beyond ability to do the required job to an acceptable level of quality, is the speed at which it will function. The maximum practical operating speed of the workhead, including the feed cycle if required, must result in a single, complete operations cycle which requires less time that the desired maximum machine cycle time. While this point may seem obvious, it is important to remember that material (work) can generally be moved as fast as one wants, and is clever enough to design for, and that controls fast enough to handle most requirements can be designed and/or selected. Thus the most prevalent cycle-time limitation is the actual performance of the work function at the work station location.

7.2.   WORK STATION REQUIREMENTS

Chapter 6 emphasized the importance of detailed specification of the machine requirement in terms of force (torque), stroke, and speed. These considerations apply primarily to basic actuators like cylinders and motors, but the same determinations must be made for off-the-shelf work stations or work station components.

Rarely is an off-the-shelf workhead such a complete package that it need only be bolted onto a table and turned on to be functional. The following considerations are essential—they fall into the realm of *design integration* as opposed to basic design.

1) *Mechanical interface* between the work station and main line work transfer subsystem. How will the workhead be mounted with respect to the transfer system so that:

    a. Work is properly oriented as it moves into the station;
    b. Work is accurately positioned with respect to workhead or tool;
    c. Work is securely held during the operation;
    d. Work departs the station and moves to next station.

2) *Speed matching* between workhead and work transfer system. Can work be moved, positioned, and clamped at the desired production rate? Can workhead be cycled during work dwell at the work station fast enough to maintain desired rate?

3) *Avoidance of machine or work damaging jam-ups* between workhead and work, adjacent workpieces, and foreign material. Can actuators be bottomed out so as to damage themselves, even as a result of control failure? Are critical clearances open to chips or dirt? It is necessary to design breakaway tooling to protect machine or work?

4) *Adjustability.* Can the workhead be adjusted—mechanically for positional accuracy; speed-wise for synchronization with transfer system; force or stroke output for fine tuning the operation on the workpiece?

5) *Access* for jam-up clearance, tool changing, troubleshooting. Is workhead accessible by operator or maintenance man?

6. *Safety* of operator, adjacent personnel. Is the active working area guarded from operator's hands during the work stroke? Can chips or workpiece fly off of machine, either during operation or transfer, and injure anyone? Are there moving machine parts which must be guarded?

Each of these considerations will be discussed in detail in Section 7.3. However, it is important to keep all of the above in mind while selecting or designing the basic workhead. Many off-the-shelf workheads were designed to be purely functional, hence none of the above factors were considered. Obviously a manufacturer of standard workheads can not be expected to make his product so that it meets all possible special machine requirements, but he would be wise to make the workhead mountable in the horizontal *or* vertical position, make it base mounted *or* able to be hung from above, etc. Workhead mounting options are an important factor in the selection of an off-the-shelf workhead.

In recognition of the fact that most users of standard workheads have unique work or product characteristics, work holding, positioning, and clamping is usually left to the special machinery designer. Although the manufacturer of the basic workhead should consider adjustability, safety, and access in the design of his product, he often does not, and these tasks are left to the machinery designer.

7.2.1. Assembly Station Requirements

Figure 2-2 depicts the critical operational requirements of an assembly station. If the assembly station could be regarded as a "black box," then one would see inputs which consisted of a workpiece and one or more parts (see Section 8.3). The output would be an assembly of these components.

To accomplish this assembly operation, the machine must first position workpiece A with respect to a machine reference MR. Fixed with respect to MR, is, of course, the workhead itself. The accuracy with which the machine must place the work with respect to MR is determined by the accuracy requirements of the assembly itself.

There are two ways in which the workpiece may be positioned within a work station. The first is *on line,* which means that the work positioning task is performed solely by the work transfer subsystem. In the case of a rotary table, the table index would place the work precisely where required to assemble and fasten part B to it. Some type of work holder, mounted on the rotary table, would probably be required to keep the work from slipping out of position during the assemble and fasten operation (as well as during indexing motion been stations).

Other on-line positioning techniques include *gating,* whereby a two-position gate stops the workpiece in its proper position directly below the workhead. The gate may simply cause the work to slide on the transfer device (e.g., a conveyor belt), or the transfer mechanism may dwell at the gated position during the assembly cycle. Note that guide bars of some type may be necessary to position the workpiece in the direction perpendicular to travel.

Another common technique is *on-line lateral clamping,* where a single clamp captures the workpiece or its work holder against a fixed stop (parallel to the work flow), or a double clamping action captures the work from both sides. This technique combines the positioning and clamping requirement. Single clamping against a fixed stop parallel to the work flow direction can result in accurate positioning (as accurate as the fixed stop), but it may not be easy to unclamp the work, free it to travel, and assure positional accuracy at a downstream station. Clamping the work from both sides alleviates the unclamping and free problem, but positional accuracy is difficult to attain with two moving parts (clamps).

Off-line lateral clamping is achieved with a long-stroke actuator (i.e., cylinder), which forces the workpiece into a very accurate positioning fixture such as a Vee-block. The work is then precisely positioned for the assembly and/or fastening operation. The problem with this technique is twofold. First, a separate actuating mechanism must be designed to move the workpiece back on line. Second, valuable machine cycle time is consumed by the off-line/on-line motion requirement.

It is also possible to physically lift the workpiece and/or work carrier as part of the main work transfer system (see Section 8.2.2). For such lift-and-carry transfer systems, the work may be very accurately located by placement into a nest from above. The primary design problem here is one of assuring overhead clearance for the transfer mechanism as it passes under a workhead (or clearance in general, depending on the machine layout).

The next critical requirement in the assembly station procedure is to feed and position part B with respect to work, A. The options for performing this step are many, and depend heavily on the relative configurations of work and part. Section 8.3 describes parts handling systems in detail, and discusses the transfer of parts to the work station. At the work station, the parts must be mated with the work. The following general options exist.

1. The part is already in position for assembly to the workpiece. Assembly motion or motions result in part placement. Examples include: automatic screwdrivers or riveters which feed parts to the driving head and hold them in jaws until they are driven into the work; automatic DIP insertion machines which hold parts (DIPs) in jaws during insertion.

2. The part is placed onto the work as the work passes by. In some cases, it is possible for the work to capture the part as it moves by on the way to the

next operation. Examples: adhesive is applied as a bead, or by a brush, as work passes by; the workpiece physically "grabs" a magazined part as it wipes by the magazine.

3. The part is placed onto a stationary workpiece by a parts placement device.

Positioning of the part B relative to the work A is essential to proper assembly. Again, the positional accuracy required depends on the relative dimensional tolerances of A and B, and the required fit between A and B. As shown above, the part positioning step may be a separate motion (which consumes machine cycle time), or it may be integrated into the fastening cycle itself (such as rivet setting).

If the part placement operation is a separate action, then the next assembly station step to be considered is that of *fastening* part B to work A.

It is not uncommon to break the assembly process down and use one station for part placement onto the workpiece, and then fasten that part at the next station. If possible, however, the part should be fastened before moving the work (and part), in order to minimize the chance for misalignment of part B to work A.

If a product is truly designed for automatic assembly (few really are) then the fastening operation will be relatively simple, that is, involve only a single machine motion. In addition, use of discrete fasteners (thus parts feeding) will be eliminated, where possible, in favor of integral fastening whereby material already existing on the workpiece is altered to fasten part to work.

Temporary fastening includes screw driving, nut running, retaining ring placement, threading, and friction insertion. With the exception of friction insertion, these types of fastening operations are the most difficult to automate because their implementation requires (a) complex motion and (b) assembly time.

Permanent fastening techniques are much easier to automate, and include: soldering, brazing, welding, adhesives, coining, staking, injected metal, ultrasonic welding (plastic), and riveting. In most cases, only a single motion is required to effect fastening. Some, including adhesive fastening, can be performed without any required motion. Adhesive can be directly applied to workpieces as they pass by an applicator.

Fastening of the applied part or parts is the last required assembly station task prior to moving the work to the next station. However, an assembly inspection task is often performed to verify the assembly. The inspection may be relatively simple, such as testing for the presence of the assembled part, or it may be complex, involving dimensional inspection of the assembly. The decision to use an inspection operation within an assembly station is usually made based on the assembly value. Very expensive assemblies should be inspected after every operation to avoid wasting downstream material and possibly precluding some cost recovery from the rejected part.

It is also recommended that assemblies be inspected, and rejected if bad, if it is probable that the bad assembly may jam the machine at a later station.

7.2.2. Inspection Station Requirements

Figure 2-3 outlines the operational requirements for an inspection station. Unlike an assembly machine, there is only one physical object (the workpiece) which must be positioned during the inspection cycle.

The important positioning requirement for an inspection station is that of the workpiece with respect to the inspection sensor. Regardless of the physical property being measured, most noncontacting electronic sensor outputs are sensitive to relative positioning of target and sensor.

Three measurement techniques are used in automatic inspection stations:

1. *Work stationary—Sensor stationary.* The work is stopped relative to a fixed sensor while the inspection measurement is made. This is the easiest method to implement, but generally only one measurement can be made per stop.

2. *Work stationary—Sensor moving.* It is sometimes necessary to scan a workpiece, taking numerous measurements at various locations over the work surface. Measurement is limited only by the speed at which the sensing device can operate. There must however, be a fixed relationship between sensor scan speed and sensor sample rate, so that measurements may be related to workpiece geometry.

3. *Work moving—Sensor stationary.* "On-the fly" inspection makes it possible to inspect a workpiece without requiring a machine cycle to do so. As the workpiece moves between stations, it can be inspected.

This type of scanning is possible if the sensor transmits and receives a signal (light beam, radio frequency electromagnetic field, microwave field, X-ray emission, etc.) which is interrupted by the traversing workpiece.

The inspection station requirement also includes that of inspection data handling. Inspection data may be used internally, i.e., only by the machine, to shut the machine down, to automatically reject out-of-spec work, or to adjust the machine so that the work remains in spec. Rarely, however, are these data used completely internally and not displayed or otherwise transmitted to the operator.

If the main purpose of the inspection station is to automatically characterize the workpiece, then data transmission and storage is essential. Usually a hard copy printout or recording is produced by the machine, which certifies the workpiece inspection.

7.2.3. Test Station Requirements

An automatic test station must perform identical positioning, measurement and data handling operations as does the inspection station (see Fig. 2-4). However, the test station must also access and actuate the workpiece so that its response may be measured.

Automatic test equipment for electronic assemblies and subassemblies is very common. Here, the actuation usually consists of application of power to various circuit points, and subsequent readout of circuit response or performance. Accessing the circuit with actuating power usually means positioning the work (circuit board) with respect to the power probes, then advancing those probes until they contact the appropriate circuit points. Often readout probes advance simultaneously with power activation probes.

Mechanical accessing of work (product) for the purpose of actuating the product is usually much more difficult. Pressure may be applied to a part, for example, by advancing a pressure hose and mask which fits the product contour and makes an O-ring seal upon contact. Automatic pressure (or vacuum) application is used to test pressure transducers, filters, pressure vessels, vacuum vessels, and to leak check instruments or devices which are to be hermetically sealed.

A complication which arises frequently with respect to test machines is that of product environment. Often it is required that the product be tested not only at ambient conditions, but also at elevated and depressed temperatures, and in various chemical or atmospheric environments. This usually involves designing an environmental chamber around the work station. The problem is then one of getting work into and out of the test chamber automatically, yet maintaining the chamber conditions.

7.2.4. Machining Station Requirements

Figure 2-6 summarizes machine tool station requirements. Positioning of work with respect to the tool is clearly of prime importance, since this defines the dimensions of the machined work. In special machinery design, the purpose of a machining station is not usually one of general machining, but one of very limited scope, such as: drill a hole, cut off a part, bend a part, etc. Since the machine will be performing repetitive operations on subsequent pieces of the identical material, tool speed is usually constant and not adjustable.

Part clamping is important during a machining operation, since forces acting on the workpiece are high. Care must be taken with automatic clamping, however, to assure that chips from previous work are not clamped with the workpiece, thereby losing accurate positioning and possibly damaging the work surface.

Chip removal is a critical design requirement. Air blasts are commonly used to clear chips, but they must be directed so that chips do not end up in critical machine clearances. Clamps or vises used to hold the work may require cleaning during machine operation, including vacuum removal of chips or even wiping with a rubber blade.

Rarely will the designer design a machining station from scratch. There are only a limited number of machining operations possible, and associated machines for performing them have been built for many years. Similarly, standard tools are almost always used (milling cutters, saws, drills, etc.) rather than specially designed tools.

The exception to the design-from-scratch rule involves dies for bending, punching, forming, or cutoff. Dies must always be designed for the specific product undergoing processing. Die design is a highly specialized field and will not be discussed in this book. Both pneumatic and hydraulic presses are available off-the-shelf for use in the special machine with a custom die set.

Automatic machining stations may also include work inspection (or inspection may be performed exclusively at the next work station) to verify the machining operation. This is particularly recommended, since subtle changes take place (tool wear, workpiece thermal expansion, etc.) which cause finished work to become progressively worse. In order to avoid producing a number of "slightly bad" (nevertheless, bad) parts, inspection is recommended.

7.2.5. Packaging Station Requirements

The general requirements for a packaging station are summarized in Fig. 2-5. As in the design of all other types of work stations, the first step is to position the work relative to the package or packaging material.

It is very often the case in packaging that the package itself is not nearly as important as the product (work). This allows a great deal of flexibility in selecting the packaging technique most amenable to automation for the particular product under consideration.

Packaging operations generally fall into two categories: (1) putting the product into the package, and (2) wrapping the package around the product. The first type of operation can further be broken down into (a) product dispensing from bulk, and (b) discrete product insertion.

Packaging product from bulk includes bottling, canning, bagging, boxing, etc., and is usually accomplished by gravity feed of premeasured amounts of product into the container. Subsequent stations close and seal the container.

Packaging discrete products into containers automatically can be done by individually stacking product into the package, or, as is more common, stacking the product, then placing the container (box) onto the stack. Again, subsequent stations take care of closing, sealing, and sorting the packages.

Wrapping stations usually require the product to be rotated along several axes relative to a stationary roll of packaging material.

Automatic package-sealing may use adhesives, hot-melt adhesive, heat sealing, or actual closure with a mating part (i.e., a cap or lid).

7.3. WORK STATION DESIGN

7.3.1. Work Positioning

The critical requirement of work positioning was shown to be the first step in all types of work station operational sequences. The ultimate object is to position the workpiece with respect to the workhead or tool, but it is more practical to reference both workpiece and tool to a fixed machine reference. Thus the workhead is positioned with respect to the machine reference when the machine is built. Each workpiece is positioned with respect to the machine reference during each machine cycle by:

1. The work transfer system alone, or
2. Clamping the work against a machine—fixed stop, or
3. Picking up the work and placing it into a precision nest.

All three of the above methods result in more or less accurate positioning of the work with respect to the machine reference and thus to the workhead.

Rotary table work transfer schemes where the work is positioned with respect to workheads solely by means of the table motion are common. Positioning accuracy is determined by table index accuracy, and in the best case may be as repeatable as ± 1 minute of arc. This translates to ± 0.010 inch at the periphery of a 6-foot-diameter table. Rotary table index accuracy can be increased by adding an index position pin which picks up precisely located holes on each table index and thereby locks the table into place. This requires the added complexity, however, of engaging and releasing the index pin for each table index.

The concept of locking the work transfer mechanism into place by engaging precision locating pins can also be used on in-line or carousel-type machines.

7.3.2. Automatic Clamping

The techniques for holding and clamping work on automatic machinery include:

1. Direct pressure pneumatic clamps,
2. Direct pressure hydraulic clamps,
3. Toggle clamp, pneumatically actuated,
4. Toggle clamp, hydraulically actuated,
5. Pneumatic or hydraulic vise,
6. Pneumatic or hydraulic chucks,
7. Pneumatic grippers (i.d. and o.d.),
8. Magnetic chucks and vises,
9. Vacuum chucks and tables.

Direct pressure clamping uses pneumatic or hydraulic cylinders to advance a clamping jaw into contact with the work, and in most cases, force the work against a fixed, precisely located machine reference. The magnitude of force application is determined by the air or oil pressure supplied to the blind end of the cylinder, and can be adjustable if a pressure regulator is used on the air or oil supply line. Clamp release may be effected by using either a single-acting spring-return cylinder or a double-acting cylinder.

Toggle clamps actuated by air or hydraulic cylinders are generally faster than direct-acting clamps because the actuating cylinder travel is much less. (Note that pneumatic clamps are always faster than the equivalent hydraulic clamps, but cannot produce the large clamping forces available with hydraulics.)

Pneumatic or hydraulic vises could be called packaged clamps because cross-slide mounted jaws are included with the actuating cylinder. Use of vises may save design time, if the vise can be conveniently integrated into the work station so that the work can flow into and out of the vise automatically. However, this is not always an easy task.

Pneumatic chucks are used to hold cylindrical work, or work with an axis of symmetry collinear with the chuck axis. Chuck jaws are usually toggle type clamps which close evenly around the rod. Rotating unions allow the chuck to be rotated without loss of clamping pressure.

Pneumatic grippers are of two types: mechanical "fingers" actuated by miniature air cylinders, or flexible pneumatic devices which expand to pick up an inside diameter. The latter type also available as o.d. grippers. The designer would not use grippers to hold work in a machine unless the grippers were also part of the work transfer system. For example, it might be desired to pick up and rotate a workpiece to several different orientations while still at a single work station. An appropriately placed i.d. gripper might effect the rotation as well as hold the work during processing.

Magnetic chucks or vises are capable of exerting high holding forces on ferromagnetic material. Magnetic chucks are amenable to high-speed automation because the holding capability (magnetic field) can be turned ON and OFF easily. Perhaps the greatest problem with magnetic chucking is that it leaves residual magnetism in the workpiece, and unless each piece is demagnetized it will pick up chips, etc. as it progresses through the machine.

Vacuum chucking is very popular in automatic machinery. Vacuum is fast acting, and can be used on any smooth nonporous surface, regardless of material. Fig. 6-40 shows the vacuum generation technique most widely used in machinery. The clamping force (vacuum) can be turned on electrically by energizing a two way solenoid valve.

7.3.3. Machine Accuracy

The accuracy with which a machine performs its required operation on the workpiece depends entirely upon the positioning accuracy of the workpiece with respect to the workhead. The following factors affect this positioning and must be taken into consideration in designing the machine.

1. *Deflection of machine members under gravity.* Workheads are often cantilevered from a central overhead support, and their supporting structures must not deflect under workhead weight.

2. *Deflection of machine members under machine loads.* Workheads performing heavy machining operations or heavy pressing operations exert large forces tending to deflect machine members.

3. *Inertial misalignments* or deflections may take place if heavy machine members move and change direction at high speed, or move in a circular path so as to cause high centrifugal forces.

4. *Impact misalignment* can occur under repeated high impact loads. Although mean machine loading may be low, impacts result in very high, albeit short-time, forces. It may not take a high speed machine long to "beat itself out of alignment".

5. *Component misalignment* with time as a result of machine vibrations. Vibrations have a tendency to loosen bolted joints, and a machine may lose its alignment accordingly. There is always a trade-off between machine adjustability and machine (long-term) stability and rigidity.

6. *Accidental overload* due to a machine jam-up. The machine structural design should take into account the possibility that an "immovable object" may become wedged between tool and work. Plastic deformation at any point in the machine structure under this condition must be avoided—either by extra heavy machine construction or by designing breakaway components.

7. *Wear* of machine bearings, cams, ways, etc. leads to misalignments and loss of accuracy over the long term. Design of a good machine lubrication system is essential. Care should also be taken to select proper wearing material combinations, and proper material hardnesses for the expected loads.

8. *Thermally induced misalignments* can be eliminated by proper design First, no loaded structural member should be positioned relative to a heat producing component (i.e., a motor) such that member sees temperature cycles of greater than 100° F, or maximum temperature of greater than 180° F. Secondly, all machine structures should be stress relieved so that operational temperatures do not result in stress relief thus deflections.

7.3.4. Adjustment Requirements

Experience has shown that no matter how carefully a machine is fabricated it can never be assembled and made to work optimally without adjustments. The reason for this is that the operating machine, under operating loads, is not the static, ideal machine which it is on paper, or even which it is as a complete assembly at rest. The complex interaction of dynamic loads, vibrations and thermal stresses on machine performance simply cannot be predicted.

The two types of adjustments which must be planned for are (a) positional and (b) timing. Positional adjustments include, but are not limited to the following:

1. Drive motor mounting adjustments to allow for belt tension, shaft alignment;
2. Drive shaft pillow block adjustment to allow for shaft alignment;
3. Cylinder mount adjustment to compensate for off-center reaction loads;
4. Cylinder stroke adjustment (or stop adjustment) to facilitate work performance;
5. Work-station-to-machine-base adjustment to facilitate accurate work positioning;
7. Work positioning stop adjustment to change machine reference;

8. Wiper adjustment for in-bowl tooling of vibrating feeders;
9. Machine leveling adjustment with respect to the floor;
10. Adjustment for interface of parts-feeding track with work transfer system;
11. Position adjustment of parts placement (pick and place) devices—both the pick-up location and the placement location.

Timing adjustments which should be considered include;

1. Work station actuation with respect to work transfer system;
2. Adjustment of dwell time with respect to index time;
3. Synchronization of workhead and work in a continuous machine.

Practical implementation of positional machine adjustment usually entails the use of shims, oversize holes, and/or slotted holes. Timing adjustments are made in the control system, and are not usually difficult unless the material transfer system and the work station are operated from the same drive shaft, with mechanical cams used as the timing devices.

Designing adjustability into a machine always compromises reliability. Bolted joints inevitably loosen under prolonged vibration. Even high-tension bolts in a static situation relax over time due to creep of the bolt material.

To the greatest extent possible, all required adjustments should be designed to be made with respect to a single, stable reference, say the machine base or table top. This allows independent adjustment of all critical components. Adjustment of two components relative to each other is accomplished by adjusting each with respect to the common reference. "Chain adjustments," where two or more critical adjustments are piggy-backed in series, generally create problems. In this case, an adjustment made to one component throws all downstream components out of alignment. Thus a single adjustment to improve machine performance may be catastrophic in its effect on other dependent machine alignments.

If adjustments must be designed into the system, then they must also be easily accessible. Not only is access essential during machine setup and debugging, but for the reasons pointed out above, it must be accessible for periodic *re*adjustment. Always remember that adjustability works both ways. By building in the ability to fine tune a machine by allowing adjustments, you also build in the opportunity to lose that adjustment.

7.3.5.   Safety Requirements

Consideration of operator safety is a very important design factor which is often overlooked or underemphasized in designing machinery. Any company or individual engaged in the business of designing and building special machinery *must* adopt and strictly adhere to the policy that:

No machine will be delivered which is not inherently safe during its operation to its operator or to those in proximity to the machine.

In addition, the machine designer and builder should make it common practice to delineate in writing, and to deliver with the machine, documents stating:

1. A detailed operating procedure for the machine;
2. Precautions against deviating from these procedures and the potential hazards of doing so;
3. Maintenance procedures and the consequences of not properly following them.
4. Any potential operator hazards which might exist, even under extraordinary circumstances (such as machine modification).

It may seem costly, perhaps irrelevent, to expect this kind of time and effort on safety, when you are usually concerned right up to the last minute as to whether or not the machine will even work. As the designer and manufacturer of the machine, however, you are responsible for anyone who is injured by your machine:

- 10, 20, or even 30 years from now;
- Whether or not the machine had been physically modified by others;
- Regardless of how ill conceived or incompetent were the intentions of the injured party;
- Whether or not operating instructions have been followed;
- Whether or not maintenance procedures have been followed.

Clearly, it is important for the design-and-build company to heed its product liability responsibilities. But what about the individual machinery designer? If there is one person who has less time to consider anything other than the functional design of the machine, it is the designer responsible for the project.

Although the individual designer may ultimately be absolved of responsibility for defective or unsafe design, it is nevertheless his professional responsibility to design the machine to be as safe as possible.

The procedure described below should be followed by the designer and others involved with the machine as an exercise specifically designated to evaluate the safety aspects of the design. The best time to carry this exercise out is early in the preliminary design phase. The overall design is well enough known to make a meaningful evaluation, yet it is not too late to make any necessary additions or modifications.

Machine Design Safety Considerations

1. *Is the machine concept inherently unsafe?* Many machines are unsafe or pose potential hazards simply because of what they do. Large, fast-moving saws, blades, high-inertia arms, and punch rams are examples.

The first question to ask is whether or not there exists a safer alternative to that proposed. If not, the risks must be minimized by protecting the operator from the hazardous conditions. Table 7-2 lists some specific safety-related design tasks.

2. *Anticipate potential accidents under assumed conditions:*

   a. Machine or machine member failure. Will pieces fly about?
   b. Improper operation. Can operator make the machine perform in an unsafe way by operating it in the manual mode?
   c. Operator panic. Can operator hurt himself if he panics and reaches into machine?
   d. Bad parts. Can defective work or parts create an unsafe condition?
   e. Overt modification. What machine elements or safety equipment might cause the operator such inconvenience that he would be tempted to remove or modify them?

3. *Is the machine fail-safe?* If the machine itself, or work therein fails, what is the most likely consequence? Can failures which might reasonably be expected be protected against?

Careful consideration of these three points should bring to light most unsafe possibilities. The designer should then keep all of these points in mind as he proceeds with the machine functional design. Such an integrated approach to machinery safety will avoid the often encountered problem of after-the-fact safety add-ons. These devices usually present major operator inconvenience, and strongly tempt the operator to remove them.

7.3.6. Reliability and Maintainability

Special purpose automatic machines are generally very expensive, and to justify this expense, many years of production by the machine are anticipated.

TABLE 7-2   Safety Related Machinery Design Tasks

- Cover all moving machine parts which could engage operator hands or feet.
- Shield work or design clamp so that an accidental unclamping of work does not allow workpiece to be propelled toward operator (or others).
- Shield operator from spray (such as coolant, wash, paint overspray, etc.) or chip ejection.
- Use additional factors of safety in securing large (high-mass), fast-moving machine parts so that failure does not result in a free-flying machine element.
- Design moving parts to fail-safe. If they do fail, make sure no one gets hurt.
- Use explosionproof motors, or fluid power, in explosive environments.
- Provide interlock controls to assure that the machine can only be operated in the sequence desired.
- Provide for automatic shutdown in the event of workpiece, part, or machine failure.
- Provide an emergency stop or panic button for operator shutdown of the machine. Place it in an easily accessible location.
- Clearly display precautions and warnings.
- Clearly display emergency procedures (use decal on the machine if necessary).
- Electrically ground machine chassis.
- Don't design safety features so they are easily removed or circumvented.

This means that the machine must be designed to be reliable, and generally, reliability cannot be assured without a good preventive maintenance program.

Reliability, like safety, is often put on the back burner during the design phase because the designer's primary concern is just getting the machine to function properly. Admittedly, reliability considerations are not glamorous—keep the base from rusting, keep the bearings lubricated, keep the electronics cool, etc. A common attitude is: "I'll worry about keeping it (the machine) working for 10 years after I get it to work at all."

There are, however, some very basic design-for-reliability considerations which cannot be attended to after the fact. The first basic consideration is that of keeping stress levels below the fatigue limit. Any machine member subject to an alternating load must be sized so that the fatigue stress endurance limit is not exceeded, taking into account all stress concentration factors. Failure to consider fatigue until after the fact will result in a major redesign effort (resizing of all structural members will basically require a new layout) or a very unsightly, unprofessional-looking beefing-up of critical members in critical areas. (Patchwork fixes on major machine designs should always be avoided—it is better in the long run to accept the economic losses associated with a redesign.)

The consequence of failing to consider fatigue stresses may be machine failure (usually cracks starting at weld fillets or internal corners) after several months or years. Usually the fix applied to such failures results in aggravating, not solving, the problem. For example, welding to fix cracks

TABLE 7-3. Some Reliability and Maintainability Design Considerations

Reliability

- Paint or otherwise coat or chemically passivate all exposed machine surfaces which may rust or corrode.
- Design all structural members to see alternating stresses less than fatigue endurance limit.
- Design all structural members to have reasonant frequencies outside the range of vibration excitation sources (motors, shafts, gear trains, high-speed impacts, etc.)
- Size ball bearings so that machine loads do not exceed bearing fatigue limits.
- Lubricate all bearings.
- Minimize bolted joints in high-vibration areas.
- Provide cooling for all electromechanical or electronic components.
- Use proper enclosures for electromechanical components for the machine environment.
- Keep chips and dirt out of bearings, side ways, and seals.
- Use off-the-shelf components when possible to take advantage of proven reliability.
- Use top-quality machine components, with warranties when possible.
- Design breakaway fixturing or tooling in high-load/high-risk areas.

Maintainability

- Provide operator access to work stations to clear jammed parts, assemblies.
- Design modular machine to aid in replacing major components.
- Provide easy access to control protective devices (fuses, breakers).
- Design machine for easy cleaning.
- Provide accessable grease, oil fittings for lubrication. Use pressurized lubrication for hard-to-access locations.

results in residual stresses, thus possible deformation and misalignment, or loss of heat treat and associated loss of parent metal strength.

Fatigue failures are hard to predict accurately, and thus to design for, primarily because of the variability of stress concentration factors. As a result, traditional machine design practice has been to err on the side of conservatism rather than to risk premature failure. Therefore, many automatic machines are somewhat heavier than perhaps they need be.

The second major reliability consideration which cannot be postponed until late in the design is that of machine vibration. The consequence of vibration are many:

- Bolted joints loosen and alignments are lost.
- High local stresses (which may lead to fatigue failure) may occur, particularly in a resonance situation.
- Vibration resulting in relatively large displacements may interfere with the machine function.

The design rule is to avoid resonances within the machine structure. This is easier said than done, however, because all machine members are complex structures having several degrees of freedom. There are potential resonant modes in torsion, in bending, and in higher harmonics of each.

Since precise vibrational analyses of machine elements are difficult, if not impossible, rules of thumb prevail and machines are designed to have very low resonant frequencies. Since heavy (high-mass) members have lower resonant frequencies, we see another motivation for designing conservatively, i.e., with heavy machine structural elements.

Table 7-3 lists some of the more common reliability and maintainability design considerations.

8. WORK TRANSFER SUBSYSTEM DESIGN

8.1. BASIC WORK FLOW CONFIGURATIONS

Section 2.4.2. pointed out that automatic machinery can be arbitrarily classified not only in terms of machine function (assembly, test, inspection, packaging, machine tool) but also in terms of the path followed by work as it moves from station to station in the machine. Typical machine types classified in this manner include:

1. In-line transfer machinery
   a. Intermittent
      i. Indexing (Fig. 8-1)
      ii. Flexible (Fig. 8-2)
   b. Continuous (Fig. 8-3)
2. Rotary (or dial) transfer machinery (Fig. 8-4)
   a. Indexing
   b. Continuous
3. Carousel transfer machinery (Fig. 8-5)   (ROt . + in-line)
   a. Indexing
   b. Continuous

The basis for classifying a machine as one of the above types is the general path followed by the workpiece or product as it moves from station to station and as sequential operations are performed on it by the machine. This progression is called *mainline* work flow, as opposed to *crossline* work flow, which defines the path followed by parts or material added to the product in an assembly or packaging machine.

More often that not, the relative velocity of workpiece with respect to workhead in a production machine will be zero at the instant the operation is performed. This period of zero relative velocity, or *dwell,* must last at least as long as the period of time required to complete the operation. In cases where the mainline transfer motion of work with respect to workhead is itself

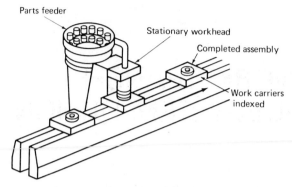

Fig. 8-1. In-line indexing machine. (Reprinted from *Automatic Assembly,* p. 13, by courtesy of Marcel Dekker, Inc.)

required to complete the operation, no dwell period need occur. Examples include: adhesive application, cleaning, machining of exterior surfaces, noncontact inspection of the workpiece, painting, slitting (e.g., roll of paper or plastic film), and environmental exposure (e.g., heat treating, drying ovens, continuous processing chambers).

If zero relative velocity between work and workhead is achieved by moving both at the same speed, then the material transfer system is said to be _continuous._ The work transfer for any of the examples above would also be continuous, because the work does not stop in order to undergo an operation. One of the problems associated with continuous machinery is that both

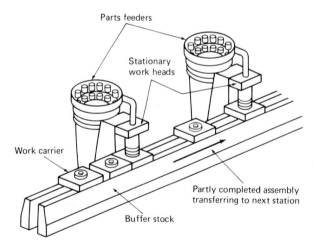

Fig. 8-2. In-line flexible machine. (Reprinted from *Automatic Assembly,* p. 25, by courtesy of Marcel Dekker, Inc.)

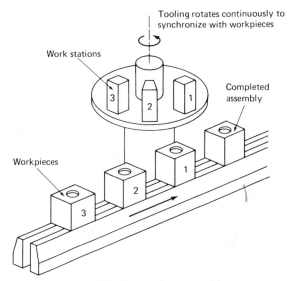

Fig. 8-3. In-line continuous machine.

workpiece and workhead are in motion—with the associated higher probability that something will go wrong. On the other hand, continuous machinery can run very fast, cycling at rates on the order of 100 pieces per minute.

If the workpiece is transferred, then stopped in order that each operation be performed, the machine is said to be *intermittent*. If the work moves from station to station at fixed time intervals, then the system is *indexing*. If the work moves from station to station at time intervals which vary depending on

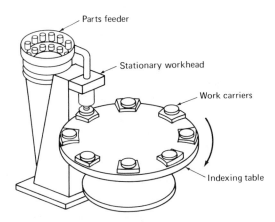

Fig. 8-4. Rotary indexing machine. (Reprinted from *Automatic Assembly,* p. 12, by courtesy of Marcel Dekker, Inc.)

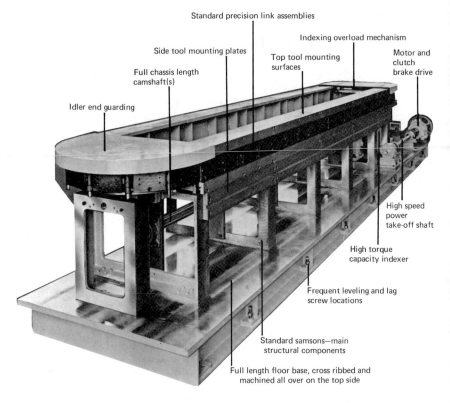

Standard precision link assemblies

Indexing overload mechanism

Side tool mounting plates

Top tool mounting surfaces

Motor and clutch brake drive

Full chassis length camshaft(s)

Idler end guarding

High speed power take-off shaft

High torque capacity indexer

Frequent leveling and lag screw locations

Standard samsons—main structural components

Full length floor base, cross ribbed and machined all over on the top side

Fig. 8-5. Carousel machine with vertical drive axis. (Courtesy Swanson-Erie Corporation, Erie, PA)

where the work is in the machine, then the transfer is *flexible* or *free*. Work transfer systems which use pallets to hold the work which have the capability of moving to the next station or stopping upon a machine control command are called *power and free* systems. They are flexible in that a workpiece may dwell at a particular work station long enough for the operation to take place, then wait in line until the next work station is free to accept it.

Flexible systems offer higher overall system reliability than do indexing machines, because a failure at one work station need not shut down the entire machine. Work can progress at all other work stations, provided there is a buffer stock of in-process work both upstream and downstream of the inoperative station. The controls for a flexible or free transfer machine are much more complex than those required for an indexing machine, as is the transfer system drive.

8.1.1. In-Line Transfer Machinery

There are three commonly used methods of moving work through an in-line machine:

1. *Pallet Transfer.* The workpiece is positioned and held in proper relation to the work stations by a pallet, or work holder. The pallet itself is then transferred through the machine by one of several possible methods, including sliding on ways, conveyor, rolling self-contained wheels along a track, or by hard attachment to a driving chain. The pallet may even have an on-board driving device. The empty work holders must be recycled back to the machine input station where a new (unprocessed) workpiece must be inserted.

2. *Non-Pallet Transfer.* In a non-pallet-type machine, the workpiece is moved through the machine without a work holder. The work is slid along ways, moved by conveyor, etc., and must usually be positioned and clamped at each work station in a separate operation so as to assure proper placement with respect to the workhead.

3. *Lift-and-Carry Transfer* (also called Walking Beam). Although this transfer method can be either pallet or non-pallet in nature, it is common enough to warrant separate attention. As the name implies, he walking beam arrangement actually lifts the workpiece from its location at one work station and transfers it to the next work station. In a multistation indexing machine, the lift-and-carry motion is usually synchronized by a rigid linkage connecting work lifters at each station. The lift-carry-place-return motion of the linkage gives rise to the term *walking beam.*

Figure 8-1 depicts the basic in-line concept, showing a machine using work carriers. The same work transfer scheme could be implemented without pallets, however, by sliding the workpieces on ways from work station to work station. Fig. 8-6 shows an in-line machine which recirculates pallets in an over-and-under configuration by dropping empty pallets after workpieces are removed at the machine output, then rapidly returning the pallet to the front end of the machine, elevating it, and finally engaging it with the in-line drive system. Fig. 8-7 shows a machine in which pallets are circulated on the same horizontal plane, thereby eliminating the added difficulty of dropping and elevating the pallets. Note however, that this scheme requires considerably more floor space than the over-and-under configuration. Both figures illustrate one inherent disadvantage of in-line, pallet-type machines—the fact

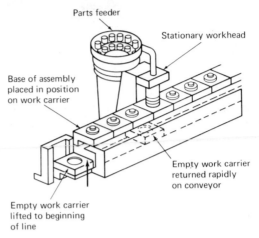

Fig. 8-6. In-line machine with over-and-under pallet return. (Reprinted from *Automatic Assembly,* p. 17, by courtesy of Marcel Dekker, Inc.)

that the pallets must somehow be moved from the output of the machine to the front end of the machine. The mechanism required to do this adds complexity and cost of the machine.

The pallet return must be rapid unless the designer is willing to allow many empty pallets. Work holders in general are expensive fixtures, precision machined to accurately locate the work with respect to the automatic workhead. The more empty pallets allowed in the return path, the more expensive the entire machine becomes.

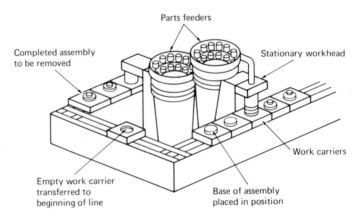

Fig. 8-7. In-line machine with pallet return in a horizontal plane. (Reprinted from *Automatic Assembly,* p. 18, by courtesy of Marcel Dekker, Inc.)

The flexible transfer concept is shown in Fig. 8.2. Since it is a basic production objective to balance work flow through a line, it would be counterproductive to design a machine which accumulated work in process at some point in the machine. Clearly, the machine must be designed so that there is one finished product at the machine output for every unfinished workpiece introduced at the machine input. It is often useful, however, to have the capability to continue running a machine in the event of a work station malfunction. As the number of work stations increases, the probability of one being down increases. If the machine were not flexible, it would be out of operation a significant portion of the time, with nothing being produced, or partially produced. By designing a capability for accumulating buffer stocks of partially finished product between work stations into the machine, the machine, can continue to operate, even though one station is down. This does not, however, solve the problem of finishing the accumulated stock of partially finished product. Once repaired, the problem work station must be run on accumulated stock during an off shift, or possibly during maintenance on the other machine work stations. If multiple machines exist, then it is possible that the partially completed product could be run through another machine (provided that machine is not being used). Regardless of the solution, it should be clear that the flexible transfer machine does indeed offer flexibility to the production engineers.

Figure 8-3 shows an in-line continuous transfer machine. The requirement of no relative motion between work and workhead is accomplished by synchronizing a turret-mounted workhead with the work motion. Multiple workheads may, however, be an expensive solution to the problem. Note also that in the example shown, that the workhead does not exactly parallel the work motion. The machine in Fig. 8-3 might, for example, be sufficient for the high-speed filling of food containers. The filling heads are relatively simple, thus inexpensive, and the slight mismatch of work path and workhead path could be tolerated. All factors considered, however, a rotary transfer system is much better suited to continuous operation than is an in-line system.

In-line work transfer systems or in-line machines offer the following advantages:

1. An unlimited number of work stations is possible.
2. All access is from one side (i.e., the front).
3. Work stations do not obstruct work, thereby making it relatively easy to unjam.
4. The machine can be placed up against a wall to conserve floor space.
5. The line can be zig-zagged as required to fit production space or work stations.

Among the disadvantages of in-line machines are:

1. Work carriers, if used, must be recycled at additional machine complexity and cost.
2. Cumulative inefficiency of machine due to many work stations increases downtime unless flexible system is used.
3. Slowest operation paces product output.

In-line assembly machines can be operated at rates as high as 3600 cycles per hour in the indexing mode, and approximately 1000 cycles per hour in the flexible mode. Lower machines rates are often more than justified, however, by the increase in machine up time.

8.1.2.   Rotary Transfer Machinery

Figure 8-4 shows a typical rotary or dial machine configuration. Work stations are situated around the table perhiphery, or, if the table is of large enough diameter, suspended inboard of the table perhiphery on a center post. At least one of the available stations must be dedicated to placing work onto the table and to removing it after processing.

Rotary machines can be either continuous or intermittent. Intermittent or indexing rotary machines are very popular because the basic table can be purchased off-the-shelf in many stock sizes and number of index locations. Design of the material transfer system is thus reduced to one of selecting the proper drive for the table, and the manufacturer's engineering staff will gladly assist in this matter. The designer is able to concentrate more time on the selection and mounting of work stations, and the design of crossline material transfer or parts feeding.

Rotary tables usually have work holders or *nests* which transport the work around the table, and provide precise work location at each station. Recycling of work holders is obviously not a problem on a rotary machine. Work carriers are not always used, however, as work can also be slid from station to station on a stationary flat table top by a rotating mask. Fig. 8-8 depicts a lift-and-carry concept applied to a rotary table.

The continuous machine design is easiest to implement in a rotary layout. Here work stations may rotate in synchronism with workpieces so as to create a zero relative velocity situation, without having to accelerate and decelerate heavy workheads. One factor to consider at high-speed operation is the effect of centrifugal forces which tend to push the work outward. On the negative side, the work must be held down to prevent its flying away. On the positive side, centrifugal force may be used to advantage in loading and unloading the machine.

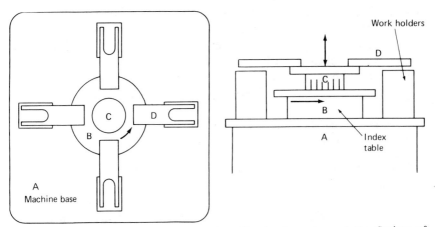

Fig. 8-8. Lift-and-carry rotary transfer device. (Reprinted courtesy of the Society of Manufacturing Engineers, Dearborn, MI)

Advantages of the rotary machine include:

1. A large selection of table sizes and index stations is available off-the-shelf.
2. Rotary tables require the minimum actual floor space. (However, they must be placed in the center of the floor.)
3. Only one work carrier per station is required.
4. Work carriers are inherently recycled.
5. High-speed continuous operation is possible.

Disadvantages of the rotary-design machine include:

1. It is limited in size (6–200 inch diameter), thus limited in number of stations. The maximum practical number of rotary table stations is 10.
2. There is limited space for work stations, hoppers, feeders, etc. at the table center.
3. Peripheral tooling, if used, obstructs access to clear jams.
4. The machine must be placed in the center of a space to permit access to clear jams, perform maintenance, thereby making inefficient use of available space.
5. It requires in-line loading and unloading in most cases, which means separate work transfer devices, mechanisms and controls.

8.1.3. Carousel Transfer Machinery

A carousel transfer machine represents a cross between an in-line machine and a rotary machine. As with rotary machines, several manufacturers supply universal carousel chassis which need only be tooled with work stations to become production machines. However, the off-the-shelf chasses are actually assembled to order (i.e., to length) by putting together standard modular components stocked by the manufacturer.

Figure 8-5 shows a typical vertical-axis carousel machine chassis. Fig. 8-9 shows an over-and-under carousel chassis. The work transfer surfaces are actual links in a large chain. Each plate can be fixtured with a work carrier, and workheads can be mounted on the stationary surfaces of the chassis.

Advantages of a carousel-type machine include:

1. An essentially unlimited number of work stations is possible, as with in-line system.
2. Work carriers are inherently recycling, as with rotary machines.
3. Carousel chasses are available more or less off-the-shelf in many sizes and with many index locations.
4. There is relatively easy access to the work to unjam the machine.

The main disadvantage is the limited space available for work stations, hoppers, and feeders mounted in the carousel center.

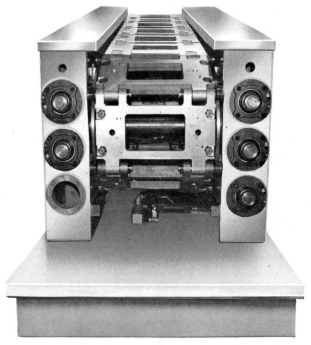

Fig. 8-9. Over-and-under carousel chassis. (Courtesy of Swanson-Erie Corporation, Erie, PA)

8.1.4. *X-Y* Table Transfer Machinery

The X-Y table as a workpiece transfer system is not used for multistation machines where one envisions the workpiece moving through the machine from input to output. Rather, X-Y tables are used predominantly for single-station, batch process machines, or for machine tool tables. In batch process applications, the machine usually consists of a work station which performs some function such as electrical testing, part insertion, etc., and an X-Y table onto which a large number of workpieces are mounted, usually manually. The workpieces are located precisely with respect to a machine reference, so that the machine "knows" the X and Y location of each piece. The X-Y table then positions each workpiece beneath the workhead so that the operation may be performed in a predetermined sequence. In this manner, the batch of work can be processed at very high rates, particularly when the table plus work has very low mass (inertia) and the distance between workpieces is small.

X-Y tables are particularly popular in the microelectronics industry, where many process steps are performed on integrated circuits while the circuits are still in slice form. There may be thousands of identical circuits on one four-inch-diameter slice, and identical processes must be performed on each. Precision X-Y tables allow the slice to be sequentially positioned such that each circuit is directly beneath the workhead, which may be a laser or a circuit test device.

On a slightly larger scale, X-Y tables are used to position circuit boards with respect to component insertion heads, so that components may be rapidly placed. There may be one or more circuit boards fixtured to the table, and each may require hundreds of components to be inserted per board. Obviously the machine must be programmed to locate the board at precisely the correct coordinates for each insertion, but once programmed, thousands of boards can be assembled very quickly.

Figure 8-10 shows an adhesive application machine, designed particularly for printed circuit board applications, which uses an X-axis drive to position the board along a programmed path while a bead of adhesive is applied to the board by an applicator head on the Y-axis drive.

X-Y tables are widely used in inspection machines to position work with respect to an inspection device, or to position or scan the device past a stationary workpiece. Features on a device being inspected can be picked up automatically by any one of a number of methods, then the distance to another feature on the piece measured by monitoring X-Y table motion required to bring the second feature to the initial reference point. Scanning devices, such as the ultrasonic "C-scan" use an X-Y drive to progressively scan 100% of the area of the work being inspected.

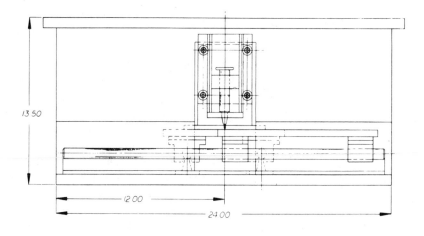

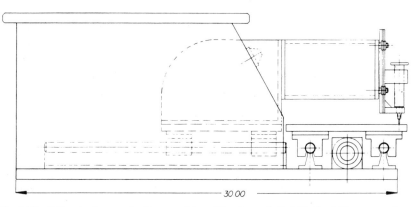

Fig. 8-10. *X-Y* adhesive application machine. (Courtesy of Knight Tool Company, Inc., Haverhill, MA)

Probably the largest use of *X-Y* (and *X-Y-Z*) tables is in machine tools. Typically, the workpiece is located with respect to the workhead for an operation to be performed, or the table drive motors actually feed the work past a cutting tool.

X-Y tables are almost exclusively driven by lead screws powered by either step motors or DC servomotors. Step motors are preferred for use with computer controls because they are essentially digital devices. Step motors do not deliver enough static holding power or torque at high speeds to be reliable for very rapid motion of heavy tables and work, so servomotors are generally used in this application.

8.1.5. Work Holders

The decision to use or not to use work holders should be made relatively early, and will certainly be presented with the concept of a proposed machine. The use of work holders, or pallets, and their proposed nature will strongly influence the basic material transfer scheme selection (in-line, rotary, carousel) as well as the overall machine cost.

It is generally preferred to avoid work holders if possible, and move the workpiece directly through the machine. Whether or not this can be done depends on the workpiece itself, the type of processing performed by the machine, and the accuracies required.

Consider first the nature of the workpiece itself. Can it be placed on or suspended from some type of track(s) so that it presents access of the working surface to the proposed work station(s)? If so, can it be slid from station to station, or must it be lifted and placed? If lifting is required, do sufficient product features exist to allow picking up (i.e., indents for a fork-type pickup, flat surface for a vacuum pickup, uniform surface for finger-type grippers)? Not only must the features for sliding or picking up be present, they must also be rugged enough to withstand the type of handling proposed.

Product which is moved without work carriers must not only be moved, but must also be positioned with respect to the work station with sufficient accuracy to allow the operation to be performed *in a highly repeatable manner*. If forces are exerted on the workpiece by the workhead, then the work must generally be clamped during the operation. Therefore, consideration of the method of work transfer must include the method of (1) positioning, and (2) clamping the product at each station. Do the features of a non-palletized workpiece allow it to slide or be placed into a positioning and clamping nest? After the operation is performed, how will the workpiece be returned to the mainline work transfer system?

Work carriers can be useful when the workpiece is complex and requires positional manipulation during the processing. The work carrier can give the product the universal handling feature which it does not itself have.

One essential feature of all work carriers, if they must be used, is that they transfer a precise reference location from the product (work) to the machine. This is the only way the workhead or tool "knows" where the workpiece is located. Since the imposition of a pallet between work and work station adds an additional tolerance buildup, it is necessary that the pallet itself be very precisely machined. As a result, pallets are expensive, and if they sustain wear or damage during service, they must be replaced.

Perhaps the most elaborate pallets are those used on a palletized flexible work transfer machine, commonly called *a power-and-free* system. The pallets either carry their own power source, or tie into a master drive system, and each pallet can be individually controlled. Common drives are:

1. Continuous chain which the pallet engages by means of a clutched sprocket. When the clutch is disengaged, the sprocket freewheels. When the clutch is engaged, the sprocket drives the pallet.

2. Continuously rotating shaft parallel to pallet motion. Pallets are driven when pallet rollers are brought into contact with the shaft. Speed can be varied by varying roller angle.

3. On-board motor which picks up low voltage power by a brush in contact with a bus bar.

4. Friction drive of pallet by belt or chain in contact with one or both sides of the pallet.

5. Pallet carried by chain-driven roller idlers moves when free, stalls with rollers idling when held.

8.2. BASIC DRIVE PRINCIPLES

The work transfer subsystem of an automatic machine usually expends its energy accelerating and decelerating the workpiece. In the case of high-inertia work moving at high intermittent rates, this power requirement can be great. The two design factors which must be determined in order to calculate the machine load due to work transfer are (1) transfer time and (2) transfer distance. Transfer time is generally the difference between machine cycle time (time per piece) and station dwell time. A factor of one-half must be applied to the transfer time if the transfer mechanism has a return stroke (as in a lift-and-carry system).

Transfer distance is usually determined by spacing of work stations, which in turn is dictated by station size. Keep in mind that station spacing will be uniform on an indexing machine, therefore minimum transfer distance will be dictated by maximum station size.

8.2.1. In-line Drive Systems

There are many possible ways to actuate in-line work transfer, the object being to move workpiece or work carrier from one station to the next. Fig. 8-11 shows a simple pawl-type drive actuated by a one-dimensional reciprocating motion. Spring-loaded pawls advance the work as the actuating piston extends to the right. As the piston returns, the pawls ride underneath the work or pallets. Care must be taken that the spring force of the pawls is not sufficient to dislocate the work on the return stroke.

Rotary motion can also be converted into one-dimensional reciprocating

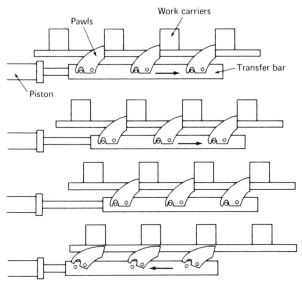

Fig. 8-11. In-line drive actuated by pawls. (Reprinted from *Automatic Assembly,* by courtesy of Marcel Dekker, Inc.)

strokes to accomplish the same purpose. Fig. 8-12(a) shows the well known crank-and-slider mechanism; in this system, however, the linear velocity is not constant. Fig. 8-12(b) shows a four-bar slider mechanism wherein proper selection of link sizes can result in constant linear velocity over the stroke for constant rotational input. Fig. 8-12(c) shows a reciprocating mechanism which converts shaft rotation into reciprocation parallel to the shaft. Note, however, that the shaft must turn at very high rpm to result in high reciprocation rates. Fig. 8-12(d) shows a skewed roller drive operating on the same principle as one of the pallet drives discussed in Section 8.1.5. Shaft rpm is constant, but linear speed can be controlled by varying the angle of contact between roller and shaft. The roller contact can be initiated by a cam fixed on the machine frame, with a reversal cam acting to reverse roller contact, and therefore slide direction, at the end of the stroke. Note that the return stroke can be made faster than the forward stroke by camming to the appropriate roller angle.

The methods of effecting one-dimensional reciprocation are too numerous to illustrate in this book. References are provided at the end of this chapter which suggest many possible mechanisms to accomplish this purpose.

Figure 8-13 shows a two-motion in-line advance mechanism which consists of first a rotation to engage pushing fingers for the advance stroke, then a reverse rotation to disengage the fingers for the return stroke. The reciprocating advance/retract stroke can be actuated by any of the previously discussed

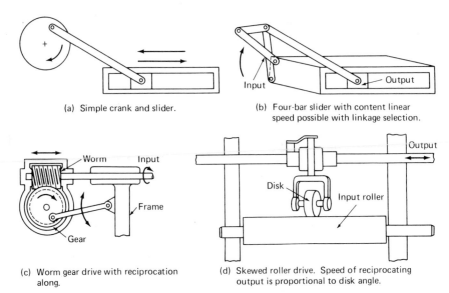

(a) Simple crank and slider.

(b) Four-bar slider with content linear speed possible with linkage selection.

(c) Worm gear drive with reciprocation along.

(d) Skewed roller drive. Speed of reciprocating output is proportional to disk angle.

Fig. 8-12. Reciprocating action from rotary drives. (Reprinted from *Mechanisms, Linkages and Mechanical Controls,* by courtesy of McGraw-Hill, Inc.)

methods. The motion used to engage/disengage pallets is here a rotation, but it could also be a linear stroke at right angles to the direction of travel which engages the pallets.

8.2.2. Lift-and-Carry Drive Systems

Many automatic machines transfer work from in-line station to station by lifting the piece, advancing it one station, depositing it at the next station, returning the transfer or lifting bar, then repeating the cycle. If one envisions a long transfer bar making an approximately circular motion while the bar remains parallel to the workpiece transfer direction, lifting all the workpieces

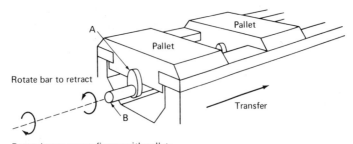

Rotate bar to engage fingers with pallets

Fig. 8-13. Two-motion in-line drive system. (Reprinted by permission of E. D. Lloyd, author of "Transfer and Unit Machines")

on the upstroke then depositing them all at the next station on the downstroke, it becomes clear why this method is often called *walking beam* transfer.

Figure 8-14 shows a lift-and-carry cycle actuated by a single reciprocating motion which could be any of those discussed in the previous section. The lifting motion is achieved by rotation of the connecting links when the transfer bar is constrained from moving to the left, and the lowering motion occurs when the transfer bar is stopped prior to the end of the stroke on the right end.

Figure 8-14(a) shows the system with the actuator fully to the left. The workpieces (or work carriers) are lifted out of their nests by the transfer bar and are ready to transferred to the right. Because the connecting links are slightly off center where they are stopped by pins, gravity holds the transfer bar plus work at the lifted position. The outward stroke of the actuator brings the transfer bar into contact with the right-hand stop as shown in Fig. 8-14(b). As the actuator bar continues to the right, the transfer bar first rises as the links rotate back over center [Figure 8-14(c)], then drop to the level shown in Fig. 8-14(d), depositing each workpiece into the next station nest. The actuator bar then returns to the left, until the transfer bar contacts the left stop [Fig. 8-14(e)], lifts the workpieces again as the transfer bar slides up against the stop and over center [Fig. 8-14(f)], then drops slightly to the initial position shown in Fig. 8-14(a).

Figure 8-15 shows a walking beam transfer system using two separate modes of actuation: one to lift and lower the transfer bar, and one to advance and retract the bar. The lifting motion could be actuated by a cam in a mechanically driven system, or by cylinders as shown in Fig. 8-15. Note that a separate degree of control must be implemented for the dual actuator scheme in order to synchronize the lift/lower and advance/retract motions. If possible it is preferable to drive the walking beam transfer system from a single actuator, using mechanical linkages to synchronize the motion. The reliability of a single-actuator mechanical-linkage system is greater than other alternatives.

Several manufacturers supply barrel-cam-driven parts handlers, often called *pick-and-place* units, which provide two-axis linear motion in the optional patterns shown in Fig. 8-16. Although primarily used for individual parts handling, these devices can be used to drive a transfer bar in a walking beam system. The combined weight of the transfer bar and workpieces must be less than the load capacity of the parts handler being considered. Fig. 8-17 shows a walking beam system driven by a parts handler. Section 8.3.5 discusses parts handlers in greater detail.

A walking beam transfer system is inherently indexing, and rather inflexible in that all work dwells at equally spaced stations for equal periods

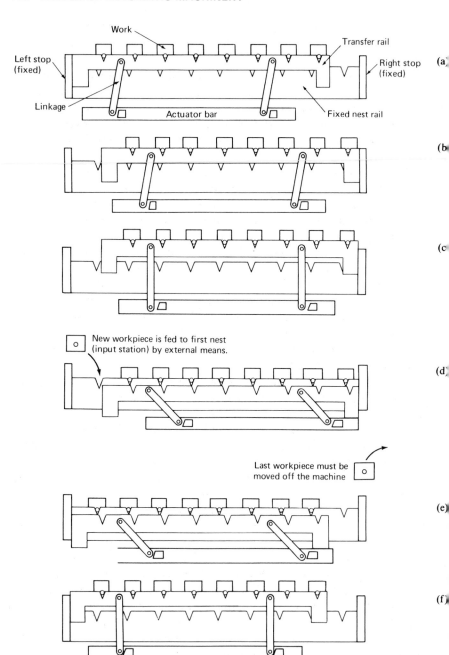

Fig. 8-14. Walking beam transfer system.

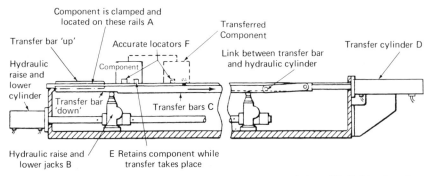

Fig. 8-15. Lift-and-carry with two actuators. (Reprinted by permission of E. D. Lloyd, author of "Transfer and Unit Machines")

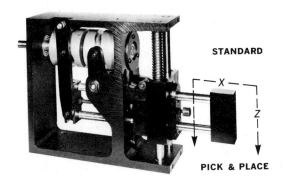

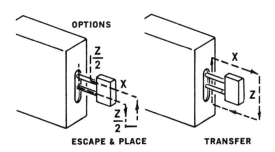

Fig. 8-16. Two-axis linear-motion parts handler. (Courtesy of Ferguson Machine Company Div. of UMC Industries, Inc., St. Louis, MO)

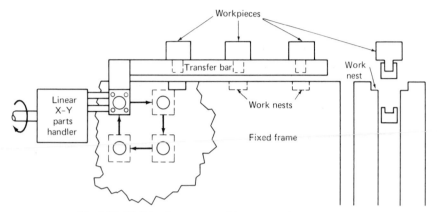

Fig. 8-17. Walking beam actuated by a two-axis parts handler.

of time. Such a system can be very fast, but care must be taken to control speed and acceleration of the transfer bar so that the work is not damaged or bounced upon pickup, and likewise upon placement at the next station. In addition, speed or direction of work transfer must not change so rapidly that the workpiece flies away. A walking beam system rarely grips the work during transfer, the weight of the work and the design of the transfer bar cradle determines how fast the machine can be cycled.

In order to maintain accurate alignment and position of the work with respect to the workhead, an accurate nest must be provided for the workpiece or for its work carrier. A nest is itself a work holder, but generally the term *nest* is asociated with a particular work station, while *work holder* is associated with a particular workpiece. One of the more demanding design jobs is that of integrating the transfer system with the work station. With a walking beam, the problem is often one of how the transfer bar lifts and places the work while avoiding interference with the nest.

8.2.3. Rotary Drive Systems

Rotary or dial-type machines rarely utilize tables designed and custom fabricated by the machine builder because such a wide range of tables and rotary drive systems are available off-the-shelf. The basic rotary table drive is sold independently of the tooling plate, which is the actual surface onto which work and associated fixturing is attached. Methods for driving the table vary, and usually reflect the manufacturer's specialty. That is, manufacturers of barrel (or crossover) cams utilize barrel-cam table drives. Cylinder manufacturers offer rotary tables driven by pneumatic or hydraulic cylinders. Since each type of rotary table has inherent advantages and disadvantages, it is wise

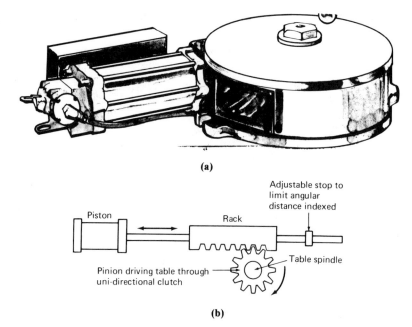

(a)

(b)

Fig. 8-18. Cylinder-actuated rotary table. (Courtesy of Schrader Bellows, Akron, OH)

to consider the table dynamics demanded by the machine before selecting the type of drive.

Figure 8-18 shows a rotary indexing table powered by a cylinder. The linear motion of the cylinder stroke is converted into rotary motion by a rack and pinion. Selection of cylinder stroke in conjunction with pinion diameter determines the angular index per stroke. In order to drive the table top or tooling plate in one direction, rather than oscillating back and forth, a one-way rotary drive mechanism must be employed, such as a wrapped spring clutch, or a ratchet and pawl. This allows the cylinder to reverse its stroke during the table dwell period as the output drive slips without affecting the table. Cylinder-powered devices which do not slip, but output an oscillating rotary motion as the cylinder extends and retracts, are called *rotary actuators*. Rotary actuators are useful in pick-and-place devices where intermittent rotary motion of less than 360° is required.

Advantages of a cylinder-driven rotary index table include the ability to implement faster retraction than extension, thus a shorter dwell time than index time. In fact, the dwell time can be varied within limits to be as short or as long as desired relative to the index time. Also, dwell time can be extended indefinitely and variably from machine cycle to machine cycle by externally controlling the cylinder fluid supply pressure.

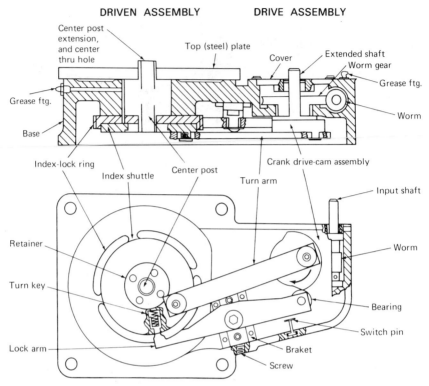

Fig. 8-19. Rotary table with worm gear drive. (Courtesy Jackson Machine Products, Madison Heights, MI)

On the negative side, the index velocity is not harmonic; rather, velocity is constant throughout the cylinder stroke. Backlash in the rack and in the clutching mechanism make this method of indexing relatively inaccurate unless a separate method of locating and locking the table is used.

Figure 8-19 shows a worm-gear rotary drive modified for harmonic indexing. Table velocity starts slow at the beginning of the index motion, increases to a maximum halfway through the index, then smoothly decelerates to the end of the index. A cam-driven lock arm accurately positions the table while disengaging the table from the drive for one revolution of the crank. The dwell time is equal to the index time in this type of table, which is relatively slow because of the worm-gear speed reduction. On the other hand, the mechanical advantage of the worm gear allows large loads to be indexed very smoothly.

Figure 8-20 shows a Geneva-mechanism-driven rotary table which, like the preceding table, features true harmonic indexing. The driver or crank is driven by a constant-velocity motor, and offers faster indexing than the worm-gear-driven table. However, the load capacity is lower because of

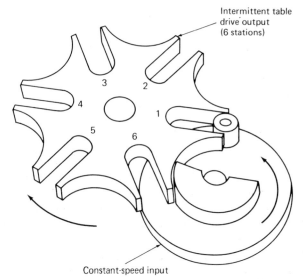

Fig. 8-20. Rotary table with geneva mechanism drive. (Courtesy of Precision Detroit Co., Inc.-Subsidiary of Fusion, Inc., Willoughby, OH)

significantly less mechanical advantage. Backlash can be considerable, so an external means of locating and holding the table at its index position should be used for applications requiring high accuracy.

Barrel or crossover cams (Fig. 8-21) are a commonly used mechanism to drive rotary tables. One revolution of the cam produces one index-and-dwell

Fig. 8-21. Rotary indexing table driven by crossover cam. (Courtesy of Swanson-Erie Corporation, Erie, PA)

cycle. The relation of dwell time to index time as well as table acceleration characteristics are made to order when the cam is cut. Index accuracy can be very good because preloaded roller cam followers are always in contact with the driving cam, so that there is no backlash. Barrel-cam rotary indexing drives are powered by constant-speed motors, and therefore are designed for use with constant-cycle indexing machinery.

8.2.4. *X-Y* Table Drive Systems

Figure 8-22 depicts the most common *X-Y* table design. The *X*-axis table is supported at the outermost edges by linear ball bearings of the open type, running on hardened steel shaftways. The linear bearings are open to clear shaftway mounting posts at intermediate points along the table travel. The table is driven by a motor-driven lead screw mounted on the moving table with fixed nut on the stationary part of the machine. This arrangement is preferred to that of a fixed lead screw with the driving nut located on the moving table because the table will not cover the motor (rendering it inaccessible) at its extreme position.

It is best to drive the table from the center, thereby counteracting the tendency of the table to rotate about its vertical axis. Supporting the table at

Fig. 8-22. *X-Y* table. (Courtesy of Power Transmission Design Magazine)

its outermost dimensions distributes the loading of the Y-axis table in such a way as to avoid lifting of one bearing when the Y-axis table is at its extreme travel. Uneven loading on the bearings results in uneven wear, thus loss of accuracy with time.

The Y-axis table should be a smaller version of the X-axis table, identical in all respects but size. It is absolutely essential that the Y-axis guideways be exactly perpendicular to the X-axis guides. Misalignment by an angle θ will result in a maximum Y-axis error of $y \cos \theta$, and a maximum X-axis error of $y \sin \theta$, where y is the maximum travel of the Y-axis table.

Step motors are the preferred drive actuators, if they are able to handle table inertia at the required indexing speeds. Very-high-speed operation with heavy table plus load often requires use of D.C. servomotors on each axis, simply because they can deliver the high torques required to accelerate and decelerate the table very quickly.

Step motor drives offer the advantage of being operable in open-loop fashion, a mode which is impossible with a servomotor. The machine controller need only supply a train of drive pulses equivalent to the distance the table is required to travel, and the table responds. If the table motion results in inertial torques reflected back to the motor shaft which exceed the step motor's holding torque capability, the table may overshoot several steps, resulting in a table positional error. Without closed-loop control, i.e., table position feedback, these overshoot errors will accumulate each time they occur, resulting in potentially large errors.

Both servomotors and step motors run in a closed-loop mode offer very accurate positioning capability. A step motor driven table can position to an accuracy of

$$x = \frac{1}{p} \cdot \frac{1}{N} \tag{8-1}$$

where: x = positional error (inches)
$\quad\quad p$ = lead screw pitch (turns/inch)
$\quad\quad N$ = number of steps per revolution of step motor.

Therefore, an X-Y table driven by a step motor with 200 steps per revolution (1.8° per step) and a lead screw with a pitch of 10 threads per inch, could position the table to within 0.0005 inch.

A DC servomotor with properly designed servosystem will position the table to an accuracy equal to the resolution of the feedback device, most likely to be a rotary digital encoder attached to the lead screw shaft. A 9-bit industrial encoder would provide a rotational accuracy of $(\frac{1}{2})^9$ revolution so, with the same 10-threads-per-inch lead screw, would provide a positional accuracy of $1/[(10)(0.002)] = 0.0002$ inch. Higher-resolution encoders are available, but then positional accuracy becomes limited by other factors, including table and servomotor dynamics, servomotor electronics and machine tolerances.

For very precise microelectronic manufacturing, X-Y table accuracies of 0.000025 inch (25 microinches) have been achieved under highly controlled environmental conditions. (Note that steel expands by about 10 microinches per inch per degree F.) Machine tool tables and industrial X-Y tables claim accuracies between 0.0001 and 0.0005 inch. X-Y tables used in DIP (Dual In-line Package) insertion machines claim accuracies of 0.001 inch over spans of two to three feet.

8.2.5. Belt and Friction Drive Systems

Belts may be used to move work either from beneath the piece as a belt conveyor or from the sides as shown in Fig. 8-23. One important design rule for using belt drives is that the positioning must not be done by the belt alone, but by an external device which traps, clamps, and releases the piece. Friction is a highly variable parameter which can change significantly because of belt stretching, wear, or contamination by liquids (water, oil, grease, etc.) common in the industrial environment.

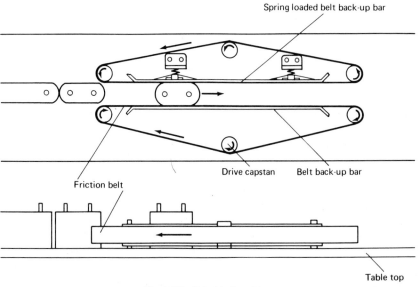

Fig. 8-23. Belt friction drive.

Friction drive systems usually have an escapement subsystem associated with each work station which (1) stops the workpiece at the proper location, (2) holds back any upstream work waiting for processing, (3) releases the piece after the operation is complete, and (4) positions the next piece for processing. The nature of the friction drive is such that simple gating of the product will hold it from advancing without significant energy losses. In other words, the driving forces are relatively low.

The friction belt drive system is clearly flexible, or nonsynchronous, as opposed to indexing. It is an inexpensive drive system, so it is amenable to branching as a means of circumventing a downed station, or as a means of sorting product (e.g., rejects versus acceptable pieces). The gating required for sorting is relatively simple, as compared with say, sorting from a walking beam system.

8.3.   PARTS HANDLING SUBSYSTEMS

The work transfer methods discussed so far in this chapter have been for mainline transfer, as opposed to crossline transfer, from the conceptual viewpoint of the generalized automatic machine model of Fig. 2-1. Inspection, test, and machine tool machinery do not as a rule require crossline material transfer. Assembly and packaging machines do require this type of transfer. Assembly machines require parts which are automatically assembled to be fed to the main workpiece. Packaging machines require containers into which product is introduced, or material which is wrapped about the product. The methods by which these additional parts or materials are introduced to and integrated with the mainline material transfer system are the subject of this section.

Many different handling solutions may be applied to any given problem. The broad design objectives are to design a system which (1) is simple, (2) is reliable, and (3) reduces handling operations. Parts hang-up or misfeeding are major reasons for automatic machinery downtime, hence the design of a reliable parts feeding system is essential to overall machine efficiency. The crossline transfer concepts listed below suggest the desired order of general handling philosophy to be applied, if possible. The concepts are presented in a descending order of overall system reliability:

(1) *Bulk Handling*. Where appropriate, feed components such as liquids, powders, adhesives directly from bulk hoppers.

(2) *Manufacture On-Line*. Stamp components directly from sheet stock, cut or form paper directly from bulk rolls, injection mold plastic parts used later on-line, form gaskets in place rather than handle individual gaskets.

3. *Manufacture Off-Line, Semi-Complete.* This is appropriate for terminals, gaskets, rubber parts, shims stamped "in-the-web" and used in rolls on machine, injection molded plastic bodies around partially stamped metal parts used as stock for further assembly.

4. *Manufacture Off-Line and Retain Orientation.* This is appropriate for magazined parts such as laminations, snap rings, integrated circuits, thin parts, and taped components such as resistors, capacitors, and screws used from rolls of taped parts.

5. *Feed Individual Parts.* This requires maximum handling:

a. Parts feeders,
b. Parts orientors,
c. Parts transfer to work station,
d. Parts escapement to work,
e. Parts placement.

Not all material can be handled as in methods 1–4. However, the designer should consider these methods to see if any are appropriate for the machine in question. Method 5 requires the most handling, but it can be adapted to handle almost any part or combination of parts over any distance and position it with respect to the automatic workhead. For this reason, such systems are widely used, and the remainder of this section concentrates on the operation, selection, and design of individual parts handling systems.

8.3.1. Parts Feeders

Parts feeders are devices which deliver parts one at a time to a point from a hopper containing a large number of randomly oriented parts. The operation of part orientation may be inherent in the feeder, or it may be due to a specially designed device which may or may not be added to the feeder. In general, the feeder itself delivers parts to a specific machine location one at a time, and those parts may or may not all be of the same orientation.

All parts feeders operate, in a sense, on gravity. That is, the effectiveness of a parts feeder depends on what orientation a part falls into when agitated. Stationary-hopper feeders rely on the probability that a parts picker will successfully retrieve a part(s) from a bowl full of randomly oriented parts. The motive power for transporting the part to a designated machine location is the picker itself, and the effect of gravity on the part.

Moving-hopper parts feeders rely on the motion which they impart to the individual parts to move the part to the designated machine location. Feeders which tumble parts rely on the fact that there is a finite probability that a randomly tumbled part will fall out of a door only large enough for one part. Obviously the probability of a sphere falling through a round hole is greater than the probability that a randomly dropped coin will fall through a rectangular slot just slightly larger in width than the coin. Vibratory hopper parts feeders operate by throwing each part ahead by a greater distance than the part slides or is thrown back again. In this manner, the parts tend to march along, and if the path is made narrow enough, they do it in single file.

The best way to feed any particular part is not always obvious, and is not a design problem which can be easily analyzed, considering the unknowns involved: the probability of parts falling in particular orientations, the interactions of parts with feeders, etc. A most useful work in this regard is *Handbook of Feeding and Orienting Techniques for Small Parts*, by Boothroyd and co-workers of the University of Massachusetts. A great deal of experimental work has gone into developing empirical feeder efficiencies versus part aspect ratio (L/D) for various generalized classes of parts as fed by different types of feeders. Data presented in the handbook allow the designer to estimate feed rates for various feeder types under consideration, and to make educated choices of the various alternatives.

Vibratory Bowl Feeders.   The most common type of parts feeder used in assembly machines is the vibratory bowl feeder. As Fig. 8-24 shows, it consists of a circular bowl with a track spiraling around the inner perhiphery

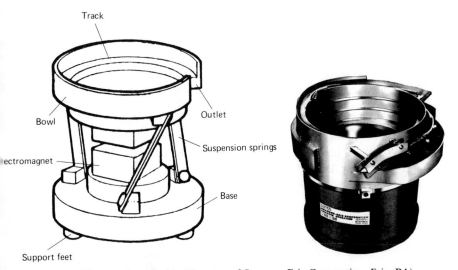

Fig. 8-24. Vibratory bowl feeder. (Courtesy of Swanson-Erie Corporation, Erie, PA)

of the bowl from bowl bottom to an outlet at the top. The feeder is excited electromagnetically, usually at 60 Hertz line frequency, although feeders are available which vibrate at other frequencies. Angled support springs result in a circumferential bowl vibration at an angle of 20–50 degrees greater than track angle. Fig. 8-25(a) depicts a part on the track of a vibratory bowl feeder. Since the angle of vibration is greater than the track angle θ the track will both lift the part and pull it forward by friction between the track and the part (on the bowl's upstroke). Forward acceleration of the bowl and part will be greatest just as the bowl starts to reverse its motion, at which point the forces on the part will be shown in Fig. 8-25(b). If the component of inertial force acting on the part is greater than the static friction force between the part and track plus the component of part weight along the track, then the part will slide up the track:*

$$\begin{bmatrix} \text{Parallel component} \\ \text{of inertial force} \end{bmatrix} \geq \begin{bmatrix} \text{Friction} \\ \text{force} \end{bmatrix} + \begin{bmatrix} \text{Parallel component} \\ \text{of weight} \end{bmatrix}$$

$$mA_0\omega^2 \cos \psi \quad \geq \quad \mu N \quad + \quad mg \sin \theta \qquad (8\text{-}2)$$

where: m = Mass of the part (lb-sec²/ft)
A_0 = Magnitude of bowl acceleration (ft/sec²)
ω = Frequency of vibration (Hz)
ψ = Angle of vibration relative to track
μ = Coefficient of friction (static) between part and track
N = Normal force between piece and track (lbs)
    = $mg \cos \theta - mA_0\omega^2 \sin \psi$ for the instant shown in Fig. 8-25(b)
g = Acceleration of gravity = 32.2 ft/sec²
θ = Track angle.

As the bowl reaches its maximum acceleration in the backward direction and just starts to move forward again, the forces acting on the part will be as shown in Fig. 8-25(c). The part will have a tendency to slide backward down the track if the inertial force and the weight of the part acting backward along the track exceed the static friction force opposing that motion:

$$\begin{bmatrix} \text{Parallel component} \\ \text{of inertial force} \\ \text{downward} \end{bmatrix} + \begin{bmatrix} \text{Parallel component} \\ \text{of weight down} \\ \text{track} \end{bmatrix} \geq \begin{bmatrix} \text{Friction} \\ \text{force} \end{bmatrix}$$

$$mA_0\omega^2 \cos \psi \quad + \quad mg \sin \theta \quad \geq \quad \mu N \qquad (8\text{-}3)$$

where: N = Normal force between piece and track
    = $mg \cos \theta + mA_0\omega^2 \sin \psi$ for the instant shown in Fig. 8-25(c).

*For a more detailed analysis of vibratory feeder dynamics, see Boothroyd, et al. (1982).

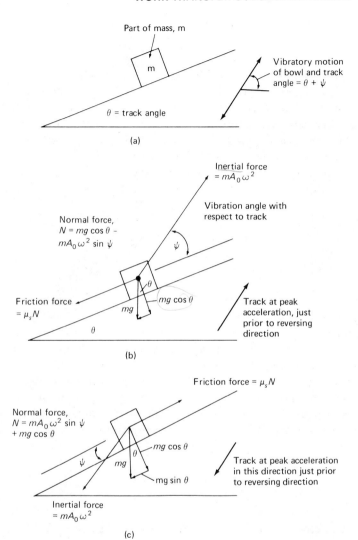

Fig. 8-25. Forces on a part on a vibratory bowl feeder track.

If Eq. (8-2) is combined with the expression for N for the case of the track accelerating forward, and rearranged, a criterion can be stated for the magnitude of bowl acceleration A_0 required for the part to slide forward:

$$\frac{A_0 \omega^2}{g} > \frac{\mu \cos \theta + \sin \theta}{\cos \psi + \mu \sin \psi}.$$ (8-4)

Similarly, the criterion for the part to slide backward at the lower point of bowl vibration can be stated as

$$\frac{A_0\omega^2}{g} > \frac{\mu \cos \theta + \sin \theta}{\cos \psi + \mu \sin \psi} \tag{8-5}$$

where the expression for N associated with Eq. (8-3) has been used.

If the threshhold acceleration for forward sliding as determined by actual values of θ, μ, and ψ, as related in Eq. (8-2) is *lower* than the threshhold for sliding backward as determined by the same values of θ, μ, and ψ as related in Eq. (8-3), then there will be *net upward sliding* of the part on the track. Inspection of the equations will show that the difference between the two threshholds at constant acceleration levels (i.e., constant $A_0\omega^2$) increases so as to increase the forward sliding tendency over the backward sliding tendency as:

1. Coefficient of friction μ increases,
2. Track angle θ decreases, and
3. Vibration angle ψ increases.

Experimental results indeed verify that feed velocity increases with the general trends described above.

It turns out, however, that the maximum possible vibratory bowl feeder velocities, and therefore feed rates, do not occur with parts in the sliding mode. As the magnitude of the bowl vibration is increased, the normal force between the part and the track on the upward vibratory stroke decreases until the part actually is thrown ahead through the air. The vibration level at which this occurs can be seen by equating the expression for normal force N associated with Eq. (8-2) to zero. The resulting criterion for part hopping is then

$$\frac{A_0\omega^2}{g} > \frac{\cos \theta}{\sin \psi} \tag{8-6}$$

A qualitative look at Eq. (8-6) shows that the tendency for the part to hop at a given level of vibration increases as:

1. Track angle θ decreases, and
2. Vibration angle ψ increases.

Analytical description of the hopping phenomenon is complex, requiring consideration not only of the point in the vibration cycle at which the part leaves the track, but also of how long the part is airborne, what the track does while the part is airborne, and the nature of the impact when the part contacts the track. Increasing the vibrational amplitude and therefore the hopping tendency has been shown to increase the part feed rate up to the point that the part bounces on impact with the track. At that point, performance is erratic, possibly because of interference among parts.

Experiment has shown the following qualitative effects to be true for parts fed by vibratory parts feeder. See Boothroyd, et. al. (1982) for quantitative details.

Effect of track vibration amplitude A_0 on parts feed rate:

1. At very small amplitudes, inertial forces on the part are unable to overcome static friction, so the part does not slide; rather, it only vibrates in synchronism with the feeder.
2. As the amplitude increases, the part slides forward, provided the criterion of Eq. (8-2) is met.
3. At a given amplitude of vibration, the feed rate is proportional to the difference between the backsliding threshold [Eq. (8-3)] and the forward sliding threshold [Eq. (8-2)]
4. When the vibrational amplitude reaches the value given by Eq. (8-6), the parts leave the track and hopping occurs.
5. Higher feed rates occur in the hopping mode as the vibrational amplitude increases.
6. Feed rate becomes erratic when vibrational amplitude is so high that parts bounce when contacting the track.

Effect of vibratory frequency ω on parts feed rate:

1. At a given vibrational amplitude $A_0\omega^2$, feed rate is inversely proportional to bowl excitation frequency, therefore lower frequencies result in higher feed rates.
2. Note that as ω decreases, the value of A_0 (vibrational amplitude) increases as required to keep $A_0\omega^2$ constant. If A_0 becomes too great, it is difficult to interface the feeder bowl parts outlet to the stationary machine.
3. A practical lower limit on vibrational frequency is 25 Hertz (25 $\sec^{-1}$)

Effect of vibration angle on parts feed rate:

1. There is an optimum vibration angle ψ for maximum possible parts feed rate, which depends on the given values of μ, θ, ω, and A_0.

Effect of track angle on parts feed rate:

1. Maximum parts feed rate occurs at zero-degree track angles.
2. As a consequence of effect 1, feed rates are greater on the flat bottom of

the bowl than on the inclined track, so that the lower parts push higher parts up the track, possibly creating jam-ups.

3. Feed rate is generally dictated by parts velocity on the bowl bottom.
4. Forward conveying is only obtained with small (less than 8°) track angles.

Effect of coefficient of friction on parts feed rate:

1. Feed velocity increases as μ increases, all other factors held constant.
2. For steel parts on steel track, $\mu = 0.2$.
3. For steel parts on rubber track, $\mu = 0.8$.

Effect of bowl load on parts feed rate:

1. As bowl empties, the maximum bowl vibrational amplitude increases (if fed with constant power), therefore feed rate increases.
2. Maximum parts feed rate occurs at approximately 25% bowl load, and decreases thereafter because pushing action of bottom parts drops off.

The design problem associated with selecting a vibtatory bowl feeder is to select one with an average unrestricted feed rate equal to or slightly exceeding the machine's part requirement. If the parts feed rate is too low, the machine will be starved for parts at the assembly work station to which the parts are being fed. If the parts feed rate is too high, the feed track supplying the work station will fill up, and the bowl feeder will simply recirculate parts up the bowl track to the bowl outlet, at which point they will fall back into the bowl. Long-term recirculation can cause wear of parts, and possibly damage.

Rotating Bowl Feeders.   Vibrating bowl feeders use hopper motion to transmit net forward motion to parts so that they walk up a spiral track. Centrifugal force keeps the parts on the track, toward the hopper wall. Rotating bowl feeders, in contrast, use bowl motion to tumble parts over a stationary exit chute. If parts orientation is necessary, then the opening to the exit chute is such that only properly oriented parts can fit through before exiting. Feed rate depends on the probability of a randomly tumbled part falling onto the exit chute, or through a selector window. Parts feed rate decreases as the number of orientation restrictions increases. Another important factor is the target (i.e., chute or opening) size relative to the part size. Openings should be as large as possible so that a properly oriented part will pass through.

Figure 8-26 shows a tumbling barrel hopper or barrel parts selector. Parts are tumbled so that they fall onto and are captured by a stationary chute

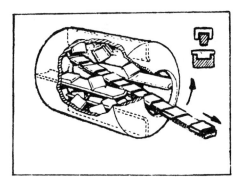

Fig. 8-26. Tumbling barrel parts feeder. (Reprinted courtesy of the Society of Manufacturing Engineers, Dearborn, MI)

running through the barrel center. Parts which do not engage the chute simply fall back into the barrel and continue to be tumbled. Note that much tumbling is inherent in the method, therefore parts which wear easily or are easily damaged should not be fed in this manner. As the inset to Fig. 8-26 demonstrates, there are many possible choices of chute geometry to catch and orient both regular and irregular parts. Parts flow out of the hopper might be by moving chute (conveyor), vibrating chute to assist parts sliding down an incline, or simply gravity alone pulling the parts down a slide.

Figure 8-27 shows a variation of a rotating hopper feeder where only the bottom of the hopper rotates, picking up parts in the proper orientation by

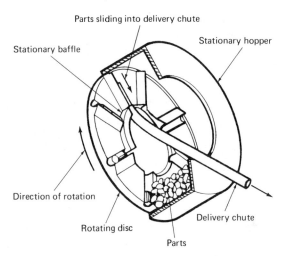

Fig. 8-27. Rotating hopper or rotary disk feeder. (Reprinted from *Automatic Assembly,* p. 73, by courtesy of Marcel Dekker, Inc.)

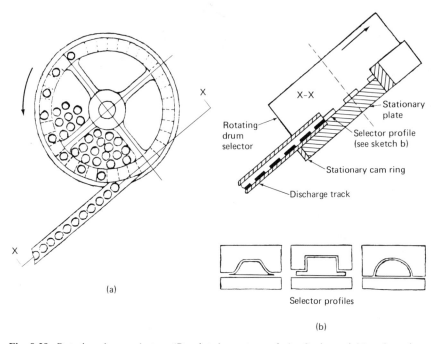

Fig. 8-28. Rotating drum selector. (Reprinted courtesy of the Society of Manufacturing Engineers, Dearborn, MI)

virtue of the way they rest in specially designed contours on the bowl bottom. As the parts are rotated to the highest point in the disk motion, parts slide into a stationary chute exiting through the bowl center. The principle is similar to the barrel hopper, but bowl contour rather than random tumbling is utilized to orient parts.

Figure 8-28 shows a rotating drum selector similar to the rotating contour disk of Fig. 8-27, except that properly oriented parts are picked up in appropriately shaped nests around the disk periphery as the disk rotates through the mass of parts. At some elevated position, away from the mass of parts, the parts drop out of their nests into a stationary chute either by gravity or by coming into contact with appropriately placed wipers. One technique is to have the nested parts drop down to a chute through an opening in the disk.

The rotating drum itself may simply pick up properly oriented parts through slots, and wipe the parts past openings in the stationary outer sleeve where they fall through to a chute. Fig. 8-29 shows an external gate hopper.

Stationary Bowl Feeders. The alternative to rotating or otherwise moving the parts hopper so that parts are transported to the exit chute in the

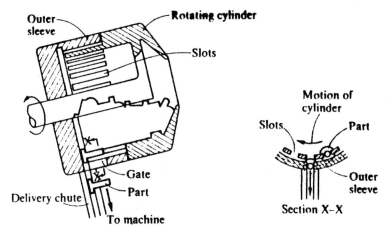

Fig. 8-29. External gate hopper. (Reprinted from *Automatic Assembly,* p. 66, by courtesy of Marcel Dekker, Inc.)

proper orientation is to move a parts picker through the mass of parts, selecting only properly oriented parts, and delivering them to a chute. In this type of feeder, the picker itself provides parts agitation, which will take place only in the region around the picker. To assist in parts agitation, vibration can be applied to the hopper.

Figure 8-30 shows a centerboard hopper selector. The centerboard dips into the parts mass, picking up properly oriented parts in a track along the top edge of the centerboard. As the centerboard swings upward and dwells at the top of the stroke, gravity slides the parts from the track into a stationary delivery chute.

Figure 8-31 shows a rotary centerboard hopper feeder, wherein the centerboard rotates rather than reciprocates. In order to allow the parts to clear the track into the exit chute, some type of intermittent drive should be employed which dwells at the point where track and chute line up.

Figure 8-32 shows a reciprocating tube hopper feeder wherein a hollow tube picks up parts as it thrusts through the parts mass. Because of the very small amount of tube-induced agitation, a separate source of agitation, say hopper vibration, will assist feed rate. Note the additional design problem requiring solution in order to transfer parts from the reciprocating tube into a stationary parts delivery chute.

Other Types of Parts Feeders. Many custom parts feeders can be designed. The object is first to obtain a one-at-a-time flow of parts from a mass of randomly oriented bulk parts. If required, that selection should be made with parts in some preferred orientation.

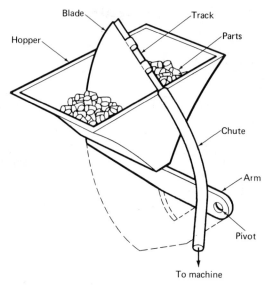

Fig. 8-30. Centerboard hopper feeder. (Reprinted from *Automatic Assembly,* p. 57, by courtesy of Marcel Dekker, Inc.)

Not all hoppers must be round. Fig. 8-33 shows an oscillating box parts feeder for feeding headed parts. The box containing the parts oscillates up and down, and only headed parts in the proper orientation can enter the exit slots.

Figure 8-34 shows an elevator-type parts selector. Properly shaped

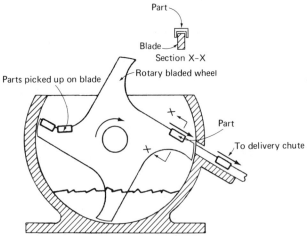

Fig. 8-31. Rotary centerboard hopper. (Reprinted from *Automatic Assembly,* p. 93, by courtesy of Marcel Dekker, Inc.)

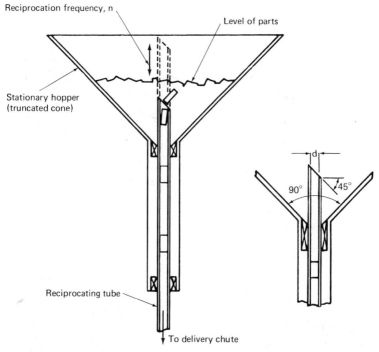

Fig. 8-32. Reciprocating-tube hopper feeder. (Reprinted from *Automatic Assembly,* p. 52, by courtesy of Marcel Dekker, Inc.)

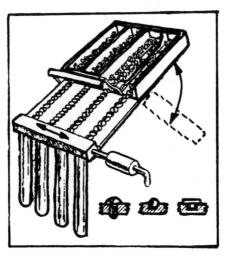

Fig. 8-33. Oscillating-box hopper. (Reprinted courtesy of the Society of Manufacturing Engineers, Dearborn, MI)

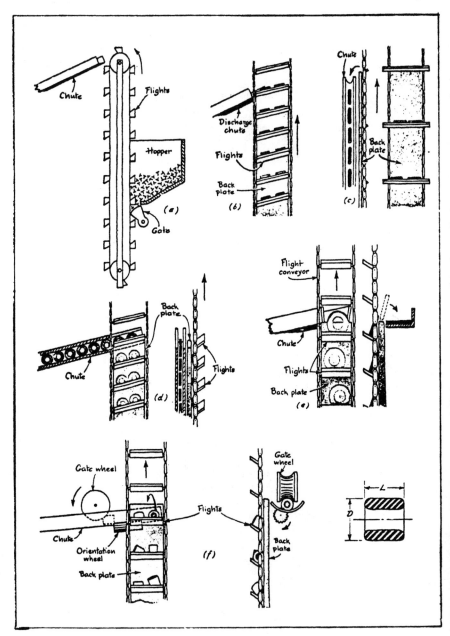

Fig. 8-34. Elevator-type feeder. (Reprinted courtesy of the Society of Manufacturing Engineers, Dearborn, MI)

contours on an endless belt circulate through the parts mass, picking up parts only in the required orientation, then dumping them into exit from a higher elevation.

8.3.2. Parts Orientors

Ideally, the parts feeder will orient the part as it selects parts one at a time from bulk, or as it delivers the part to the machine location where it is to be used. Usually, however, a custom orienting device will have to be designed to handle the particular part.

Parts orientation should be performed as close to the parts mass as is practical. Ideally, the part would be picked in the proper orientation and delivered directly and positively to the exit chute. In this manner, excess wear resulting from parts recirculation would be avoided. If the method of rejecting improperly oriented parts is used, as it most commonly is, then the rejection should be such that rejected parts fall back into the parts hopper, where they can recirculate. Parts orientation external to the feeder hopper should be avoided unless a method is used which assures 100% proper orientation further down the line, prior to parts use in the machine. In other words, every part which leaves the hopper should be used by the machine in its assembly operation.

Notice that all of the parts feeders other than the vibratory bowl feeder discussed in the previous section perform the orientation function by virtue of their unique design. Orientation is assured by one of the following methods:

1. Agitate or tumble parts so that some number fall onto an exit chute which will only accept properly oriented parts. Parts are then ready to be transferred directly to the machine work station. Feed rate depends on the probability of a randomly oriented part being captured by the exit chute.

2. Agitate or tumble parts so that only properly oriented parts pass through a custom-shaped opening or gate into the appropriately positioned exit track or chute.

3. Pass a picker through the parts mass which, by virtue of its design, can only pick up parts in the proper orientation. If possible, the picker should be integral with the exit chute. If not, the picker must transfer the part in the proper orientation onto the exit chute.

Vibratory bowl feeders, which are the most widely used hopper-type parts feeders, use *in-bowl tooling* to orient parts which travel up the spiral feeder

Wiper rejects more than one thickness of parts, returning them to the bowl. Contours may be used to pass given profiles only.

Spacer forces parts with head up over the track edge past their center of gravity, returning them to the bowl.

Air jet wiper blows improperly oriented parts back into the bowl.

Cutout passes parts with sharp, open end down, but lets rounded cup end fall back into the bowl.

  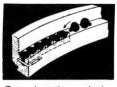

Cutout with a hood holds parts standing up as they pass over gap, while parts with horizontal axis drop back into bowl.

Rails over cutout retain parts with head up, permit parts with head down to drop back into the bowl.

Two orientation methods for use with long, headed parts. Slot lets parts hang by heads. At discharge tube, arm trips parts to fall head first if desired.

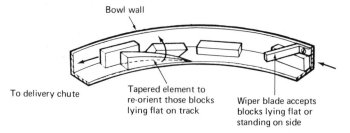

Bowl wall

To delivery chute

Tapered element to re-orient those blocks lying flat on track

Wiper blade accepts blocks lying flat or standing on side

Fig. 8-35. In-bowl parts orientation. (Reprinted courtesy of the Society of Manufacturing Engineers, Dearborn, MI)

track. The tooling is generally located at the top of the bowl, near the parts outlet, and positioned so that improperly oriented parts fall back into the hopper.

Figure 8-35 shows a number of techniques for in-bowl orientation of parts being fed by vibratory bowl feeders. Notice that the most common technique is rejection, whereby improperly oriented parts are knocked back into the bowl as they pass wipers or over cutouts. It is also possible in some cases to reorient parts by tailoring track geometry so that the part changes orientation as it progresses forward.

A general design procedure for in-bowl orientor tooling includes the following steps:

1. Determine the number of possible orientations of the part as it is fed up the track. Keep in mind that randomly agitated parts will seek minimum potential energy positions, which generally means that the part center of gravity will be a minimum distance above the track. Recall, however, that there are potential energy plateaus in some orientations which require an input energy level greater than the bowl feeder is capable of delivering to allow the part to reach the next lower potential energy state. Thus, one would not expect cylinders with L/D ratios of five to be fed up a track on end. However, a cylinder with L/D of 1.5 may well feed up the track on end, even though the minimum potential energy position would be on its side.

2. Provide a device to reposition parts into a more favorable orientation if possible. Reorientation as opposed to part rejection will maximize parts feed rate. Thus rectangular parts on edge may be nudged over to lie flat, without knocking the part back into the bowl.

3. For each remaining undesirable orientation, provide a wiper or a cutout to reject parts in that orientation.

4. Make sure rejected parts are recycled by rejecting them back into the feeder bowl.

5. Avoid wipers which present an abrupt edge to parts in such a way that several parts may wedge and lock under the wiper. For parts having a tendency to jam together, use a wiper which is angled with respect to the part motion so that it gently nudges tangled parts back into the bowl. Break all sharp edges on wiper parts.

6. Design adjustments into wipers to allow for fine tuning during the machine debugging process. Slotted holes are common.

7. If necessary, consider active orientors such as air blasts or moving wipers.

8.3.3. Parts Transfer

Once parts leave the parts feeder, presumably in proper orientation, they must then be transported to the work station, where they are mated to the workpiece in an automatic assembly operation. The most common method of transport is the gravity feed track. Fig. 8-36 shows a typical open slide track for transporting headed parts, here a transistor can. Feed tracks may also be power assisted, by air or by vibratory excitation. The powered type usually

Fig. 8-36. Open-slide parts transfer track.

still use gravity as the primary moving force, with air assist for higher speed operation, or vibration to minimize parts hang-up or friction.

A gravity assist track must be as loose as possible to facilitate sliding, but not so loose as to allow the parts to lose orientation as they are transported. Open tracks are preferred to caged or completely enclosed tracks because parts jam-ups can be readily fixed.

The following track design rules should be followed to the greatest extent possible in designing a feed track for a particular part.

1. Make tracks as simple as possible. Flat ways, or open contour ways are preferred over partially or fully enclosed ways. Minimize the number of points at which the part contacts the track.

2. In a gravity-powered track, the track angle must be greater than:

$$\theta \text{ (radians)} > \tan^{-1} \mu \qquad (8\text{-}7)$$

where μ = static coefficient of friction. This is the minimum angle at which gravity will overcome static friction, and cause the part to slide. If headed parts are being fed, it is possible that too great a track angle can result in the part tipping and hanging up as shown in Fig. 8-37. Boothroyd, et al. (1982) show that the maximum track angle θ which can be tolerated before the situation shown in Fig. 8-37 occurs is

$$\theta < \tan^{-1}\left[\frac{\mu D(\mu d + 2l + h) - \mu h \cos \alpha(2l + h)}{\cos d\,[\mu d(2l + h) - h^2 - d^2] + hD}\right] \qquad (8\text{-}8)$$

where: μ = static coefficient of friction
$D = d \sin \alpha + h \cos \alpha$ (inches)
d = diameter of the head (inches)
l = distance from the bottom of the head to the center of gravity of the part (inches)
h = height of the head (inches).

3. Minimize friction between the part and the track:

    a. Select track material with the lowest possible μ with respect to the part material.

    b. Tracks should be as hard as possible to avoid scratches, dents, etc., which would increase the effective μ.

    c. Break sharp edges on tracks with stone or paper. Tracks should be as smooth as possible, possibly polished.

    d. If necessary, coat the ways with low-μ agent, e.g., Teflon.

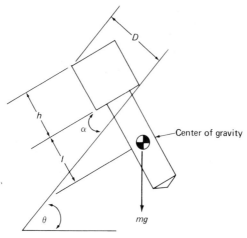

Fig. 8-37. Parts jam in feed track due to cocking.

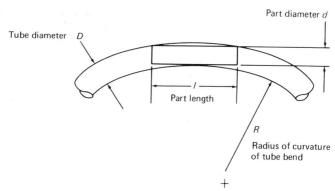

Fig. 8-38. Minimum radius of curvature for a tube track.

4. Minimize track length. The feeder bowl should be as close as possible to the work station consistent with access to the work station and maximum allowable track angle.

5. Maximize radii in track bends. Too sharp bends in a track may result in blockage. Consider the cylindrical part in a tubular feed track, as shown in Fig. 8-38. Boothroyd, et al. (1982) show that the minimum tube diameter D required to allow the part to pass through the tube is

$$D = 0.5 \{[(R + d)^2 + (l/2)^2]^{1/2} - (R - d)\} \qquad (8\text{-}9)$$

where D = Minimum tube diameter (inches)
  R = Radius of the feed tube bend (inches)
  d = Diameter of the part (inches)
  l = Length of the part (inches).

The behavior of parts passing through the feed track radius should be carefully thought out, as this is the location at which problems (hang-ups) are most likely to occur. Consider the effects of subsequent parts pushing from behind as will occur when the feed track fills up.

6. Maximize accessibility to the parts on the track to allow an operator to quickly fix any parts jams. Use enclosures to maintain parts orientation only if absolutely necessary.

7. Use active assists if necessary. Air assist jets may be used with tubular feed tracks as shown in Fig. 8-39. The air assist works best when parts are transferred one at a time, however, as the air flow ahead of the part creates a partial vacuum by virtue of its motion. Individual parts fed one at a time can

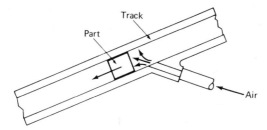

Fig. 8-39. Air-assist tubular feed track.

be transported to the workhead very quickly by the combined action of upstream vacuum and downstream pressure.

Parts transported by feed tracks usually stack up at the bottom of the track because they are fed at a faster rate than the machine uses them. This is prudent practice, to assure that the machine workhead is never starved for parts. The parts are then gated one at a time from the feed track backlog to the workstation by means of an *escapement*.

8.3.4. Escapements

An escapement mechanism is one which releases or "escapes" one part at a time from a backlog of parts, all similarly oriented, and backed up in a single line. The parts backlog may be on feed tracks (discussed in the previous section) or in a preloaded magazine. Escapements may also be used to release fixed quantities of bulk material (i.e., liquids, powders) rather than discrete parts.

Escapement should take place as near to the workhead as possible. Ideally, escapement would take place with a short, positive motion which leaves the part in place for a subsequent workhead assembly operation. Mechanisms which accomplish this are the subject of this section. Mechanisms which must pick individual parts from a queue and place them at another location fall into the category of parts placement devices, the subject of the following section.

Positive escapement is the process by which a part is removed from the queue and pushed or guided to its proper workhead location by a single machine stroke. The part is effectively placed where it belongs in the work station. Passive escapement is the process of simply escaping or gating the part, then allowing it to fall or slide into place under its own weight. There is a place for both types, with positive escapement being preferred with smaller parts and higher-speed machines. The objective of positive escapement is to avoid parts misfeed and to assure accurate, repeatable position of a part relative to its workpiece.

Escapement of a part so that it is properly positioned on the workpiece is very important, but not often easy to implement. One complete machine station may have to be devoted to simply escaping and placing one part onto an assembly. This is usually required when the escapement and placement of a part must occur from the same position relative to the work as does the work station function. If the escapement mechanism can place the part and move out of the way, then only one work station and one index cycle need be used. If the two mechanisms interfere, then separate stations may be required.

One possibility for avoiding separate part placing stations is to place the part "on the fly" as the assembly is approaching the work station. If the "part" is soldering flux or adhesive, this can be done. It would be harder to place screws into threaded holes as the workpiece moves by.

The unique feature of an escapement mechanism is that it releases one part at a time, while holding back the backlogged parts line. The procedure may require several discrete steps which must be performed in the proper sequence, as follows:

1. Capture and hold the second part in line.
2. Release the first part in line.
3. Position the second part as the new "first" in line.
4. Repeat cycle.

Alternatively, the escapement mechanism may simply wipe the first part in line away, following the wiping motion abruptly with a barrier which blocks the second part in line from advancing. The problem with this method is that the second part is still trying to advance. The abrupt barrier, as a minimum consequence, wears the part; in the worst case, it jams with the part caught trying to advance.

The six basic classifications of escapements are: (1) ratchet, (2) slide, (3) rotary, (4) gate, (5) displacement, and (6) jaw. Brief descriptions of each follow.

Ratchet Escapement. Fig. 8-40(a) shows a simple pawl escapement of parts in a track. The pawl motion is easily implemented by a rotating cam, which rocks the pawl back and forth once each machine cycle. The cam action may well be taken off the main machine drive shaft, thereby assuring synchronization with the machine.

Figure 8-40(b) shows a ratchet escapement actuated by a single reciprocation such as that obtained by a cylinder or a solenoid. Fig. 8-40(c) shows the same concept, except that two actuators are required, and they must be synchronized so that the upper ratchet captures its part slightly before the first part is released.

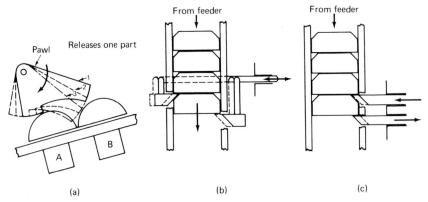

Fig. 8-40. Ratchet escapement. (Reprinted from *Automatic Assembly,* by courtesy of Marcel Dekker, Inc.)

Care must be taken in using a ratchet that the ratchet smoothly engage the held part without transmitting a force back through the parts, tending to move the line backward. Also, the pawl should not damage (i.e., scrape or score) the part.

Ratchet escapements release parts in a straight line. Also, they do not themselves act in a positive manner unless additional mechanism is added.

Slide Escapements.   Slide escapements change the direction of motion of the part being fed. In general, the part drops into a cavity in or just ahead of the slide with the part itself holding back subsequent parts. As depicted in Fig. 8-41(a), a transverse motion of the slide moves the part to its proper location, or to another chute. The slide motion must be quick, so that the next part does not interfere with the slide and jam the mechanism. As shown in Fig. 8-41(b), a single slide can escape several feed chutes simultaneously.

Slide escapements can be used with parts which do not interlock with each other, slide relatively easily with respect to adjacent parts, and are not damaged by the wiping of a previous part away and the subsequent impact of the slide preventing its forward motion. The slide clearances must be close, so that the slide does not partially capture and jam the next part. The slide should be made of a hard, wear-resistant material and should have rounded, burr-free edges on its parts-pushing surface.

As in the case of ratchet escapements, care should be taken not to allow the parts line to advance then be forced back as the slide passes. Such motion can cause jam-ups far back up the line.

In the case of a magazine feeder, the parts, usually flat, are stacked and spring loaded in a magazine. Slide escapements are appropriate for magazine feeders. One variation on the slide is the *rotating friction wheel escapement,*

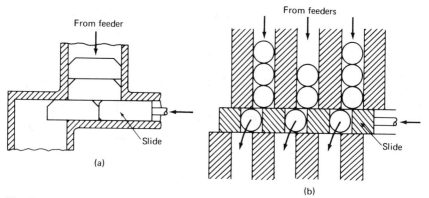

Fig. 8-41. Slide escapements. (Reprinted from *Automatic Assembly,* by courtesy of Marcel Dekker, Inc.)

shown in Fig. 8-42. As parts are required, the friction wheel rotates one revolution, partially feeding out one part. The part is then picked up by subsequent parts handlers and pulled completely out of the magazine.

(3) **Rotary Escapements.** If the slider of a slide escapement becomes a circular component driven by continuous or intermittent rotary motion, then the escapement is of the rotary type.

Figure 8-43(a) shows a simple turnstile escapement. Fig. 8-43(b) shows a drum escapement commonly used for escaping cylindrical objects, such as cigarettes. The drum escapement can also be used to dispense bulk materials, as shown in Fig. 8-43(c). Fig. 8-43(d) shows a star-wheel or drum-spider esapement which dispenses flat, cylindrical objects. The rotating axis may be either horizontal or vertical; however, if it is horizontal, the parts must be spring loaded as in a magazine.

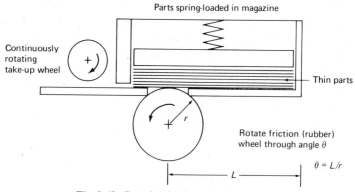

Fig. 8-42. Rotating friction wheel escapement.

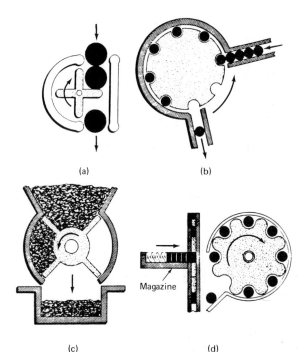

Fig. 8-43. Rotary escapements. (Reprinted courtesy of the Society of Manufacturing Engineers, Dearborn, MI)

4. Gate Escapements. A gate escapement is in essence a simplified ratchet. The gate simply opens and closes, rather than capturing the next part before releasing the first part. Gate escapements are often conditional devices, in that they open or divert a part in a particular direction, depending on whether a specific condition has been fulfilled. The condition is recognized by a mechanical, electrical, or pneumatic sensor. The machine control system then decides on appropriate gate action and provides the gate with the required actuating signal.

Gate escapements are common on conveyors and other transfer devices where parts move horizontally and are spaced rather than tightly queued.

5. Displacement Escapements. A displacement-type escapement is a mechanism by which a fixed volume of bulk material is released, or escaped, as it is displaced by an actuator. As such, this type of escapement is used mostly to dispense liquids and thixotropic materials such as adhesives, rubber, plastics, etc.

A displacement actuator is usually a plunger or piston which moves a fixed distance, thereby dispensing a fixed volume of material. The syringe of Fig. 6-

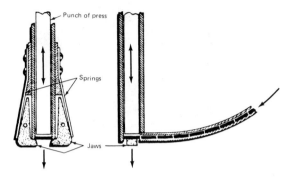

Fig. 8-44. Jaw escapement. (Reprinted courtesy of the Society of Manufacturing Engineers, Dearborn, MI)

14 used to apply adhesive to a workpiece would be a displacement-type escapement. The plunger would be moved a fixed distance each machine cycle, thereby dispensing a fixed volume of adhesive. The surface tension of the adhesive will prevent dripping between cycles.

Jaw Escapements. Jaw-type escapements are used to hold parts in position for the actual work stroke to occur. Fig. 8-44 shows a jaw escapement holding rivets for a staking operation. The jaws hold the rivet in its proper location (presumably the mainline work transfer system has positioned the work directly below). The rivet setter drives the rivet out of engagement with the jaws, pushing the spring-loaded jaws out of the way in the process.

Jaw escapements are used for accurate parts location, and are usually used with automatic screwdrivers and component insertion machines.

Escapement Control. Escapement control is a very important machine design consideration. The escapement must cycle at the same speed as does the associated workhead. Since the part must usually be in place before a workhead operation takes place, the escapement actuation control signal must lead the workhead control signal by a constant time. This period will depend on how long it takes the part, once escaped, to reach its proper position.

If parts can be escaped "on the fly," that is, the workpiece can effectively pick up a properly placed part as it moves by the escapement, then control of the escapement is implemented by the workpiece.

If a separate work station and therefore a separate index cycle is required for part placement, the escapement mechanism in effect becomes a work station, and its control signal is applied during the workpiece dwell at that station. If part placement and subsequent operation take place at the same station, during the same machine cycle, then control signal synchronization becomes critical. In very-high-speed machinery, the part may be on its way out of the queue before the workpiece has arrived at the station. Controls must be timed so that part and work arrive at the proper place at the same instant.

The control scheme for an escapement mechanism and placement operation can be very basic:

1. Escape part.
2. Perform operation.

It can also vary in complexity up to the following extreme:

1. Check for part in escapement.
   a. If none, deactivate escapement.
   b. Next, deactivate work station.
   c. Next, signal operator.
2. Simultaneously perform step 1 for presence of workpiece.
3. Positively place part onto workpiece.
4. Check for proper part placement on workpiece.
   a. If wrong, disable work station.
   b. Signal operator.
5. Perform assembly operation (i.e., actuate workhead).
6. Move completed assembly to next work station.

8.3.5. Parts Placement

If the escapement does not itself actively position the part required for assembly to the workpiece, then a separate parts placement device may be necessary. In many cases, the parts placement device will replace the escapement, picking parts from a transfer line and placing them onto the workpiece. Parts placement devices have an additional advantage over escapements in that they can change the orientation of a part from the way it has been transferred along feed tracks to that required by the assembly. This allows the parts to be transferred in the most efficient, reliable orientation regardless of orientation required at the workhead.

Parts placement devices are appropriately called *pick-and-place parts handlers,* or, as has become fashionable recently, *pick-and-place robots.*

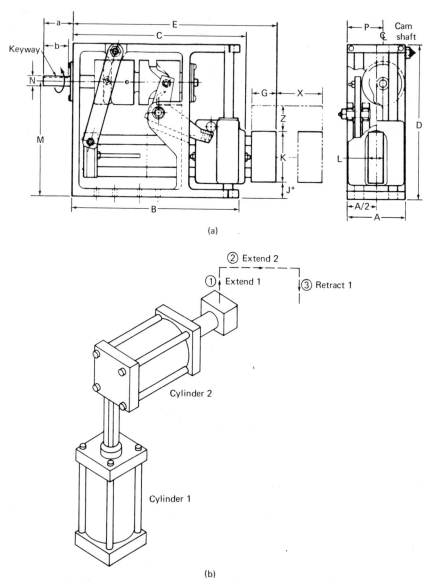

(a)

(b)

Fig. 8-45. Linear pick-and-place parts handlers. ((a) Courtesy of Ferguson Machine Company Div: of UMC Industries, Inc., St. Louis, MO)

Such units are available off-the-shelf with many pick-and-place motions and lengths of stroke. The designer would most probably be able to use a standard pick-and-place unit for all parts handling requirements. Only under the most unique of machine requirements would one attempt to design his own device.

Figure 8-45(a) shows a linear-motion pick-and-place device driven by a

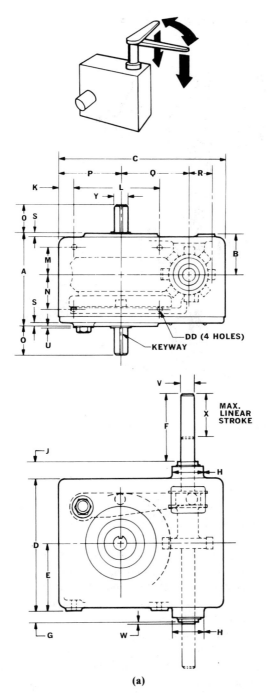

Fig. 8-46. Rotary pick-and-place handlers. (Courtesy of Ferguson Machine Company Div. of UMC Industries, Inc., St. Louis, MO)

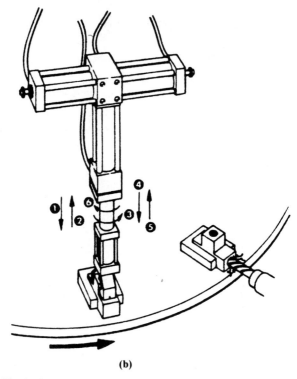

(b)

Fig. 8-46. (*Continued*) (Courtesy of PMD Inc., Ft. Wayne, IN)

barrel-cam mechanism. Fig. 8-45(b) shows a linear pick-and-place motion implemented by two pneumatic double-acting cylinders.

Figure 8-46(a) shows a rotary pick-and-place parts handler which uses a barrel-cam actuator, and Fig. 8-46(b) shows a rotary device which operates by means of double-acting cylinders.

Both the linear and rotary barrel-cam-type parts handlers can be made to order by the manufacturer for special applications. Stroke lengths, sequences, dwell times, and accelerations can all be controlled by appropriate cam design. Custom-designed cams are expensive, and once the cam is cut, the handler performance is fixed.

Cam-actuated handlers are designed for constant-rpm drives; their application is therefore limited to indexing-type machinery. Cylinder-operated placement devices can be controlled for cycles of varying length, or for varying dwell times during a single cycle. For this reason, they are applicable in flexible machinery.

Selecting the proper parts handling device is only one-third of the design job. Tooling to pick up individual parts must be designed, and sequencing controls to engage and disengage the pick-up action must be provided.

The three most common tooling methods used to pick parts are:

1. *Grippers.* Grippers are usually mechanical linkages operated by small pneumatic cylinders, or inflatable plastic fingers which grip inside or outside dimensions when inflated. Air is commonly used because the gripping components are light-weight. Mechanical grippers could be actuated by electrical solenoids, at some increase in weight.

2. *Vacuum pickup.* Vacuum pickup is common for handling flat, delicate parts. The part must have a surface against which the vacuum head can seal.

3. *Magnetic pickup.* An electromagnet is heavier than the two previous tools, but can be controlled (turned on and off) more rapidly. The parts must be ferromagnetic in order to be picked, and should not be so light that residual magnetism holds them even when the magnet is turned off.

Parts placement devices can be used effectively to actuate mainline work transfer as well as in crossline parts feeding applications. If workpieces are light enough, the linear pick-and-place transfer device may be used to drive a lift-and-carry work transfer rack.

REFERENCES

Boothroyd, G.; Poli, C.; Murch, L.; *AUTOMATIC ASSEMBLY,* 1982, Marcel Dekker, Inc., NY.

Boyes, William E.; *JIGS AND FIXTURES,* 2nd Edition, 1982, Society of Manufacturing Engineers, Dearborn, MI.

Chironis, Nicholas P. (Ed.): *MECHANISMS, LINKAGES AND MECHANICAL CONTROLS,* 1966, McGraw Hill Book Company, NY.

Lloyd, E.D.; *TRANSFER AND UNIT MACHINES,* Industrial Press, Inc. NY.

Moskalenko, V.A.; *MECHANISMS,* 1964, Hayden Book Co., Inc., NY.

Treer, Kenneth R. (Ed.); *AUTOMATED ASSEMBLY,* 1979, Society of Manufacturing Engineers, Dearborn, MI.

9. MACHINE CONTROL SYSTEM

Reference to Fig. 2.7, reminds us that the total machine design process consists of:

1. The selection and/or design of the machine *work stations* (Chapters 6,7);
2. The selection and/or design of the machine *material transfer subsystem* (Chapter 8);
3. The design of the machine *control system.*

This chapter discusses design techniques and off-the-shelf hardware commonly used to control automatic production machinery. Chapter 10 presents the various sensors commonly used with machine control systems.

9.1. CONTROL SYSTEM REQUIREMENTS

9.1.1. Types of Manufacturing Processes

Production processes can be grouped into three general categories:

1. *Continuous processes.* These are processes wherein there is only one workpiece from raw material input to finished product output. Various individual operations take place on the workpiece as it moves through the process facility, and at any one time all operations are being simultaneously performed at different workpiece locations. Examples include: paper processing, sheet metal production, newspaper printing, fabric dying and print, fiber production, many chemical processes, sewage treatment, etc.

Control of these processes is usually referred to as *process control* rather than *machine control,* and the control problem is one of *maintaining* conditions (temperature, humidity, pressure, tension, etc.) at a *set point* for each workpiece location as required by the process.

2) *Batch processes.* These are processes, usually chemical as opposed to mechanical, whereby measured amounts (batches) of raw materials are processed. The entire batch undergoes processing in a manner not unlike the continuous process above. Examples include: many food production processes, paint manufacture, custom plastics blending, casting operations, etc.

The control problem in batch processing is one of attaining and maintaining process conditions for each batch, *and* of sequencing batch operations.

3. *Discrete component manufacture.* Individual workpieces are handled and processed according to a predetermined sequence. Examples include: automobiles, machined metal parts, computers, etc. All assembled products fall into this category.

The control problem is generally one of *sequencing* discrete operations and workpiece motions by turning actuators on and off at the appropriate times.

9.1.2. Automatic Machinery Controls—Sequencing

The type of control of interest with respect to automatic machinery is that of discrete component manufacture. As such, the primary control problem is one of sequencing machine operations. In its simplest form, sequencing consists of switching various active machine elements (motors, cylinders, solenoids) ON and OFF in the correct order and for the correct periods of time.

If we are designing an automatic machine, by definition the machine should sense or otherwise determine the successful completion of one operation and automatically initiate (switch on) the next operation. Alternatively, if the completion of a machine operation is determined (sensed) by the operator, who in turn initiates the next operation, e.g., by actuating a switch, then the machine is said to be *operator paced,* and the machine is *semiautomatic* at best.

There are two alternatives to fully automating a sequence of machine operations. The first consists of switching various active elements ON and OFF strictly as a function of time. Indeed, many older machines have implemented sequence control with an electromechanical drum timer, shown in Figs. 9-1 and 9-2. This type of control device is also called a *program timer.*

The drum is driven at a constant speed by an AC synchronous motor. Cams located around the drum periphery determine when, during a single drum revolution, a particular switch will be ON and OFF. Each machine element to be turned ON and OFF during each cycle requires its own switch, and its own ring of cams at a particular axial location on the drum. The angular position, circumferential extent, and number of cams at a particular axial location

Fig. 9-1. Cam-actuated program timer. (Courtesy of Precision Timer Co., Inc., Westbrook, CT)

determine when during the cycle, how long during the cycle, and how many times during the cycle (i.e., one complete drum revolution) each switch, thus active machine element, will be ON.

Note that this type of sequence controller can be used to cycle the controlled machine at any speed (assuming appropriate drive motor controls)

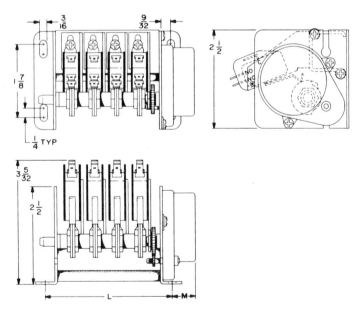

Fig. 9-2. Schematic of timer of Fig. 9-1.

up to drive motor synchronous speed. In addition, if the circumferential location of cams can be changed, the sequence controller is *programmable* in that relative timing of switch closures within a single drum rotation can be changed at will.

Since there are no *sensors* or *feedback* associated with this controller it cannot operate with any knowledge of actual machine operations. The machine thus runs in *open loop* fashion, with active machine elements being switched ON and OFF as a function of time only.

The second alternative to automating machine sequences is *event* based. Here, completion of a machine operation must be sensed by an event *sensor* (e.g., limit switch, proximity switch, pressure switch, etc.) and the output of that sensor used to initiate the next machine operation. The manner in which the machine control system uses actual machine event information to implement subsequent machine operations is called the *control logic*.

Consider the drum sequencer of Fig. 9-1 with the drive (timing) motor replaced by a step motor. The drum indexes one angular step for each excitation of the step motor, and closes those switches for which cams have been positioned. Further, assume that the signal to actuate the step motor is initiated by a sensor located somewhere on the machine in response to a particular machine event. The machine will then perform sequential operations, not as a function of time, but as a function of completion of the previous operation or other machine event.

A control device similar in concept to the stepped drum sequencer just described is a *stepping switch*. An event-based signal causes the stepping switch to advance one step, thereby opening (or closing) a different set of multiple switch contacts. The construction is somewhat different also from that described for an electromechanical drum sequencer.

9.1.3. Machine Control Requirements

In general, the control system of an automatic machine performs three basic functions: control, protection, and monitoring. The *control* function is that which assures proper operational sequencing and synchronization of all machine subsystems as the machine cycles. This function is necessary in all machine control systems.

The *protection* function is that which assures that neither machine nor operator are damaged as the result of an abnormal condition. Such controls are either *interlocks,* which prevent the machine from operating unless a specified combination of safe conditions exist, or *cutouts,* which stop machine operation when an unsafe condition is detected. It is highly recommended, but not absolutely necessary, that the control system perform this function.

The *monitoring* function of the control system is necessary for all but the simplest open-loop systems. Monitoring requires sensors to detect machine condition(s) or performance and then uses this information *internally* to implement the control or protection functions or *externally* to advise, alert, or inform the operator. It is also quite common for machine information to be transmitted externally to a recording device, a storage device, or even a higher-level computer capable of implementing on-line changes in the process or machine operation.

Figure 9-3 shows a generalized machine control system. The basic requirements of the control system are to start and continue the sequential and synchronized operations of the work transfer subsystem and the work stations, until stopped by the operator or other control condition. The most common control systems are event based, in that completion of one operation or motion sets the next operation or motion into action (hence the need for work station and work transfer sensors). In the event of in-process inspection of the work, a sensor is required to make that inspection. This inspection may be required to trigger subsequent operations (i.e., continue if part or assembly is within specification), or the inspection data may simply be stored as production control records. The former case represents a control function, while the latter case illustrates a pure monitoring function.

Basic control functions which require operator interfacing include START, STOP, EMERGENCY STOP, and JOG. JOG is a control function which allows the operator to move the machine through its sequential cycle manually (i.e., by pushing the JOG button). This function is not necessary to the automatic operation of the machine, but it is invaluable in debugging or troubleshooting the machine, clearing out machine jam-ups, and starting up the machine after shutdown.

Notice in Fig. 9-3 that a single *controller* box accepts all sensor and operator control inputs and implements appropriate control outputs. Obviously the controller is a complex system in itself, and indeed is the subject of most of this chapter. This controller might be comprised of a bank of relays, a hard-wired solid state electronic control system, a bank of pneumatic or hydraulic valves and other fluid components, a commercially available programmable controller, a custom-made or commercially available microcomputer, or a relatively large commercial minicomputer. In most cases, however, a device or subsystem will be required (*input signal conditioner* to transform the sensor output signal into a signal compatible with the controller. Likewise, an *output signal conditioner* device or subsystem is usually required to transform the controller output signal into a signal capable of actuating the appropriate work station or work transfer system.

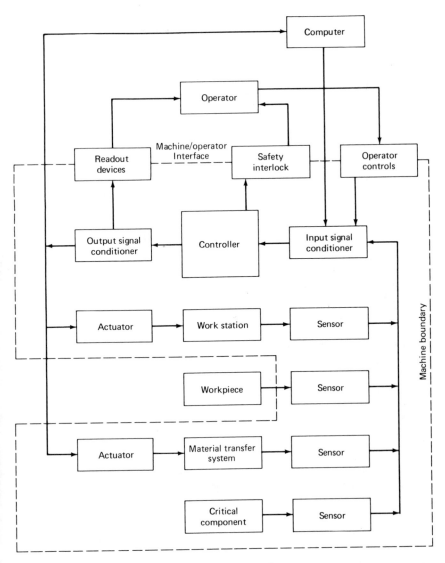

Fig. 9-3. Generalized control system.

Table 9-1 summarizes the most common machine control requirements encountered in the design of automatic machinery. It is important to select those control functions and requirements for the machine early in the design phase and to avoid add-on requirements late in the design or after the machine is built. If control requirements cannot be fully defined early in the

TABLE 9-1.   Machine Control Requirements.

Operator interface
ON/OFF
MANUAL/AUTOMATIC mode selection
JOG (for setup, unjamming, debugging)
EMERGENCY STOP (panic button)

Sequence specification (if machine is programmable)

Data or status displays
On-board diagnostics (for troubleshooting)

Safety interlocks (Machine will not start until operator assumes safe position, etc.)
Safety cutouts (Machine shuts down if operator assumes unsafe position, e.g., puts hand
    through a light screen)

Internal control
Operational sequencing
Speed controls
    Maintain fixed speed
    Assume speeds at various discrete levels
    Assume continuously variable speed
    Forward/Reverse
Acceleration controls

Initializing machine (upon startup)
Internal calibration

Monitoring
    Data acquisition
    Data manipulation
    Data storage
    In-process inspection (Use results to automatically reject bad work)
    Status monitoring (Use results to stop machine)
      If jam-up detected
      If potential jam-up detected (e.g., bad part or assembly)
      If machine out of specification (e.g., overheats)

---

project, then the controller box shown in Fig. 9-3 must definitely be of the
programmable type, with sufficient input and output capability to accommodate all subsequent additions.

9.1.4. The Control Scheme

Concept design of a machine control system is not unlike the concept design
of the machine itself, in that the first step requires the designer to prepare a
concise listing of *just what the control system must do.* This detailed
definition of what the control system must do so that the machine performs as
desired is called the *control scheme.*

Design of the control scheme does not require in-depth knowledge of automatic control technology, electronics, or computers. *It simply requires a clear understanding of what you want the machine to do.* Thus there is no inherent training or experience deficiency which prevents even the novice machine designer from fully defining the control scheme.

Figure 9-4 shows one possible control scheme is list form for the machine for which requirements are defined in Fig. 4-4. Notice that the control scheme can be defined even without selection of the specific control components or machine components. It is simply a statement of what the control system must do. The scheme of Fig. 9-4 is very basic, in that it makes the machine perform only required operations on the work. It does not include protective functions as discussed in 9.1.3., or monitoring functions other than those required to trigger subsequent machine operations.

---

1. Input X-Y coordinates to controller memory:

   Hole 1, Hole 2, Hole 3, Start of milling traverse, End of milling traverse.

2. Input z coordinates for drilling feeds:

   Center drilling; Hole 1, 3 drilling; Hole 2 drilling; Hole 1, 3 tapping; Hole 2 counter-bore.

3. All tools will be set up manually in correct turret order at correct tool positions.

4. There will be three control modes, selected manually by operator prior to starting machine:

   Automatic—Machine will perform one full cycle then return to block input position, unclamp finished block, await unload and load of new blank.

   Fast Cycle—Machine will operate as in automatic cycle as long as JOG button is held ON.

   Slow Cycle—Machine will operate at $\frac{1}{4}$ speed as long as JOG button is held ON.

5. X-Y table and drill feed position will be continuously monitored with appropriate sensors.

6. Sequence control will be event based, with subsequent operations performed only after X-Y table or drill feed coordinates of previous operation are achieved. (See Fig. 4-4 for process operation details).

7. Upon the START command, X-Y table will home to a precision hard stop position, initializing table sensors to (0, 0).

Fig. 9-4. Control scheme for a machining station.

The first step in developing a control scheme is to specify basic control requirements, much as was done in Fig. 9-4. The second step in control scheme development is to include machine and operator protection functions, again in qualitative terms. The third step would include similar consideration of all desired monitoring functions.

Several examples of possible additions to the basic control scheme of Fig. 9-4 follow. Specific functions depend on customer requirements, local electrical codes, workpiece value, machine value, operator injury potential, etc.

- Determine whether or not block is indeed in fixture. If not, turn machine off (*protection*). Signal operator and display problem (*monitoring*).
- Sense dull tools or broken tools by magnitude of torque at specific work positions (*monitoring*), use logic to ascertain that a problem exists and shut down machine (*protection*).
- Display process information to operator or transmit to data storage (*monitoring*).
- Separate circuit breakers on drill heads, table drive motors (*protection*). Indicate specific motor for which breaker is open (*monitoring*).

Recall that the control scheme is a concise listing of control system requirements without consideration for how each requirement will be implemented. With reference to Fig. 9-3, the control scheme would specify which work stations and work transfer actuators would be energized according to what sequence. Although it would not necessarily identify specific sensors, it would indicate the need for sensors to perform the desired control, protection, and monitoring functions.

A specific control system design procedure will be developed in a later section. Such subjects as selection of controller technology and detailed control system design can be better addressed after general consideration of the various machine control technologies.

9.1.5.   Control Power Levels

Control is the *direction* and *timing* of high-level power outputs using low-level power inputs. For most machinery control systems, high-level power outputs include:

- Line voltage (440, 220, 110 V AC) to power motors, solenoids, lights, or heaters;
- Shop air pressure (80–100 psig) to operate pneumatic cylinders;
- Hydraulic pressure and flow (2000–5000 psig over a very wide range of flow rates) to operate hydraulic cylinders.

It is generally not feasible or safe to switch high-level power directly in response to machine operation. It is common practice to implement machine sensing and control logic at low power levels, then amplify the low-level control signals (control outputs) to levels appropriate to operate the work station and work transfer system actuators. The amplification and whatever other signal modification might be necessary falls into the functional box labelled "control output signal conditioning" in Fig. 9-3.

Typical low-level control power inputs used to direct high-level power outputs include:

1. Manual button pushing (ounces of force);
2. Transformer-reduced AC voltage (20 V AC);
3. DC rectified from line voltage (12–28 V DC) or batteries;
4. Sensor outputs (2–20 mA, 0–10 V DC, millivolts DC);
5. Fluidic air pressure levels (5–15 psig);
6. Logic-level electrical inputs (0–5 DC for TTL, 0–15 V DC for CMOS).

9.2. RELAY CONTROL SYSTEMS

9.2.1. General

Machine control as an engineering discipline has evolved over the years, just as have most other areas of engineering. However, there is probably no better example of historical influence over modern applications of new technology than in this specific area. Microprocessor-based programmable controllers (PC's) utilize integrated circuit logic to implement machine control, however, PC's are typically programmed in relay ladder logic because this terminology and symbology is well understood by electricians, plant engineers, and machine maintenance personnel.

Although use of control relays in machine control systems is on the decline, relay logic continues to provide practical insight into the design of a machine control system. This practical understanding of just what the machine is being instructed to do is not so evident in more abstract control system design techniques utilizing truth tables, Boolean algebra, or microprocessor programming.

Relays were originated for use in telegraph transmission. The telegraph sounder is itself a relay—when the coil is energized, an armature moves, resulting in a clicking sound. As telegraph lines became longer (i.e., out of the laboratory and into the field), transmission losses due to wire resistance became a problem. It was impossible to flow sufficient current through a very long (high-resistance) wire to activate a coil at the other end. Use of intermediate coils which closed contacts to energize the downstream circuits reduced effective line resistance by one-half.

Since this intermediate coil and contact system repeated the action of the sender key, it relayed the signal down the line. Hence the name *relay*. Other historical (pre–solid state) uses of relays to implement logic on a large scale included:

1. Railroads—traffic controls and switching;
2. Telephones;
3. Traffic control lights;
4. Machinery controls.

9.2.2. Electromechanical Relay Construction

A relay is an electromechanical device which actuates one or more switch contact closures whenever power is applied to the relay *coil*. The relay coil is simply a low power electromagnet which, when actuated, moves a mechanical armature which actuates contact closure (or opening). Fig. 9-5 shows a typical control relay. Fig. 9-6 schematically shows relay operation.

A *control relay* is particularly useful because actuation of the relay coil results in the closure of several (up to 12) contact pairs. If the contact pairs are

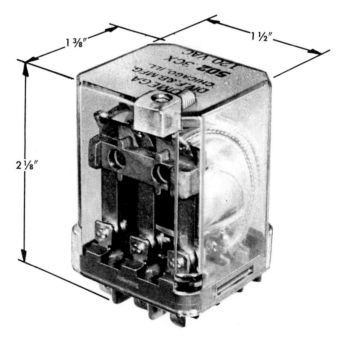

Fig. 9-5. Standard industrial control relay. (Courtesy of Omega Electric, Chicago, IL)

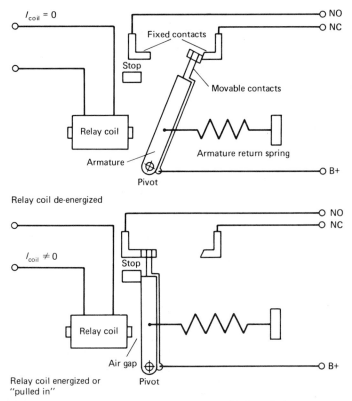

When relay coil is energized by applying voltage to the coil, the coil electromagnet pulls the armature up against a stop, maintaining an air gap between the armature and the coil core. The NO contacts close, passing current between the NO terminal and B+. The NC contacts which were originally passing current, open, no longer passing current. When the coil is de-energized, the armature return spring returns the armature and contacts to their original position.

Fig. 9-6. Electromagnetic relay schematic.

normally open (NO) then they will close, resulting in a short circuit, when the coil is energized. If the contact pairs are normally closed (NC), then they open when the coil is energized. In general, the coil and contacts of a control relay operate at the same level of voltage, but the contacts are capable of conducting relatively high currents (up to 10 amps). As defined in NEMA standards (Industrial Control ICS-1970) the term *relay* does not refer to devices controlling high power loads. The maximum relay load includes motors and solenoids drawing 2 amps.

A *contactor* is used to switch high power levels (say 440 V AC, up to 100 amps) as might be required by large motors or heaters. The coil energizing power required is relatively low and is carried out at control voltage levels. A relay used in this context is an *amplifier*, in that low-level power is used to switch high-level power ON and OFF.

A *time-delay* relay is one in which the contacts do not react immediately to relay coil energization. *Time-delay-on-operate* relays exhibit contact closure (or opening) after a preset time delay following coil energization. The contacts remain in their switched position as long as the coil power is present.

Time-delay-on-release relays transfer contacts immediately upon coil energization. Upon removing power from the coil, the time delay occurs before the contacts transfer back to their non-actuated position.

Time delay relays are available with both fixed and adjustable time-delay periods. Delay times ranging from several seconds up to one hour are possible. Longer time delays are normally implemented with a timer (see section 10.1.6).

In addition to armature actuated control and power relays, there are also now available solid state relays (SSR's), reed relays, and mercury contact relays.

Reed relays have relatively low power contacts (1 or 2 amps) but can be actuated by logic-level coil power, that is, directly by the output of a computer or other logic device.

Mercury contact relays offer the advantage of long contact life by virtue of the liquid mercury which makes and breaks each contact.

Solid state relays are used for very high-speed switching applications where electromechanical relay contacts would rapidly wear out. In addition, SSR's can be energized with logic-level power, so that they may be driven directly by a computer.

Table 9-2 summarizes most of the many types of relays available on the market today. For purposes of basic relay logic discussion, only the electromechanical control relay, the power relay, and time delay relays need to be considered. Practical control system design may well dictate consideration of other relay types available.

9.2.3. Relay Ladder Logic

Relay ladder logic is a method of implementing sequential machine control by means of control relays. Fig. 9-7 summarizes typical symbols used to depict loads, relay coils, relay contacts, manual switch contacts, pressure switch contacts, and limit switch contacts. Refer to Section 10.2 for limit switch fundamentals. These symbols are used to construct a *relay ladder diagram* which not only shows the physical interconnection of control components, but also shows the sequence of events which the control system will implement.

TABLE 9-2. Relay Characteristics.

Type	Typical Life (cycles)	Operate Time (msec)	Release Time (msec)	Coil Operate Voltage	Maximum Number Contacts	Maximum Contact Current (amps)	
						break	continuous
NEMA Size 1							
motor				110–600	3 hi-amp		
contactor	200K	40	55	V AC	4 lo-amp	15	27
General purpose							
industrial				110–600	6 NO		
relay	200K	40	40	V AC/DC	6 NC	6	10
Small							
industrial				6–220	12 NO		
relay	1M	2	5	V AC/DC	12 NC	5	10
Reed relay	2M	15	5	150 V Max. AC/DC	12	3	5

Rules for constructing a relay ladder diagram follow:

1. Vertical sides of the ladder diagram represent control power and ground. The line on the left is the "hot" line (also called B+) and the ground line is on the right. Usually AC control voltage is somewhat reduced from line voltage, so the vertical ladder sides are shown as stepped down from line power by a transformer.

Note that the insertion of a conductor between the sides of the relay ladder will cause current to flow through that conductor.

2. To construct a control circuit, start at the top of the ladder and work down. Strings of control components between sides of the ladder simulate rungs of the ladder diagram.

3. To the greatest extent possible, draw circuit "rungs" in the same order as they will be energized in the control sequence. (This is not always possible, since the sequence often backtracks up the ladder, particularly where a repetition of a control action takes place within a complete cycle.)

4. Contacts of a control relay carry the same designation as the relay coil which energizes them. (1-CR, 2-CR, etc.) Different relays (coils) are designated by the number preceding CR.

5. Multiple contacts actuated by the same coil are lettered to designate the different contact pairs (1-CR-A, 1-CR-B, etc.)

6. Draw all switch and relay contacts in their "normal" or non-actuated state.

Exception: Often a machine in its OFF or "home" position results in a cam holding a normally open (NO) limit switch closed or a normally closed (NC) limit switch open. These switches are diagrammed in their held positions with an arrow apparently holding the switch in its "abnormal" state. See Figure 9-7 for proper symbol.

7. All switch terminals should be designated as normally open (NO) or normally closed (NC) on the diagram.

The relay ladder diagram for a machine with limit switches should be accompanied with a separate pictorial diagram showing the actuation of limit switches by various controlled machine actuators.

8. All switching should be done on the "hot" side of the line. (Relay contacts, limit switches, timer contacts, etc.)

The ground line of the ladder should take return current from all load devices. (Solenoids, relay coils, lights, etc.)

This convention minimizes the chances of getting two load devices (incorrectly) connected in series under certain switch conditions.

Consider now some basic control circuits using the relay ladder diagram technique. Fig. 9-8 illustrates a very common relay control circuit known as a *holding* or *stick* circuit. Note that the line voltage is transformed down to, say, 120 V AC. Performing the required control switching at lower voltages results in less arcing when contacts are opened, thus longer contact life. The motor, running at line voltage, is connected across the line and is switched on by 1-M, a *motor starter* or *contactor* which has heavy-duty contacts rated for the higher power levels.

The diagram is read from top to bottom as follows. When momentary contact manual push button 1-PB is pushed, current flows from B+ to ground, energizing motor contactor coil 1-M and red light R, indicating that the system is ON. NO contacts 1-M-A close, thereby creating a current path through 2-PB, 1-M-A, and 1-M. This results in coil 1-M remaining energized or sticking ON, even though momentary contact 1-PB is released.

Power to coil 1-M closes the high-voltage contacts 1-M-B, 1-M-C, and 1-M-D supplying power directly to the motor.

The motor will run until the STOP push button 2-PB is pushed. Current can no longer flow to coil 1-M, so the armature drops out. Contacts 1-M-A, 1-M-B, 1-M-C, and 1-M-D also return to their de-energized, NO state. Power is no longer supplied to the motor, which therefore stops. Note that when 2-PB is released, current can no longer flow to coil 1-M because contacts 1-M-A

Symbol	Description
	Load device — solenoid coil, brake, clutch
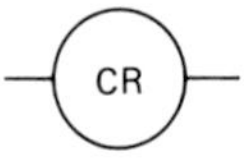 CR	Control relay — Number before CR, i.e., 1-CR, 2-CR, etc., identifies relay coil.
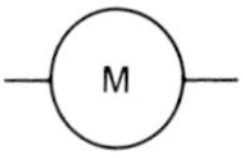 M	Motor contactor (starter) — Higher current rated contacts than in CR, numbers identify each coil, i.e., 1-M, 2-M, etc.
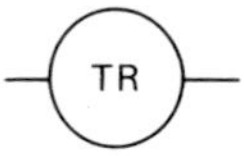 TR	Time delay relay coil — Number identifies coil, i.e., 1-TR, 2-TR, etc.
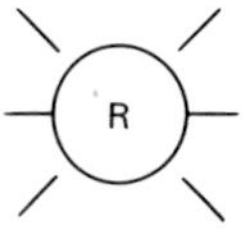 R	Light — R = red, B = blue, W = white, etc.
Relay Contacts	
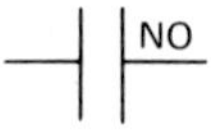 NO	Normally open contacts (NO) — Also called Form A contacts. Letter after coil designation identifies contact, i.e., 1-CR-A, 1-CR-B, etc.
 NC	Normally closed contacts (NC) — Also called Form B contacts. Letter designates contact, i.e., 2-CR-A, 2-CR-B, etc.
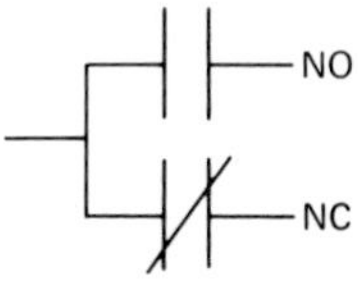 NO — NC	Transfer contacts (NO, NC) — Also called Form C or "break-before-make" contacts. Both contacts carry the same designation, i.e., 3-CR-A, 3-CR-B, etc.
 NO	Normally open (NO) time-delay relay contact. Time delay is after coil is energized.
 NC	Normally closed (NC) time-delay relay contacts. Time delay is after coil is energized.
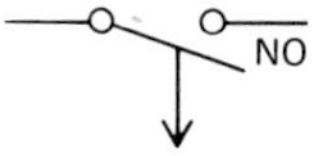 NO	NO time-delay relay contact. Time delay is after coil is de-energized.

Fig. 9-7. Typical relay ladder diagram symbols.

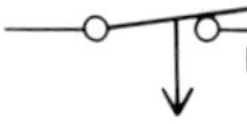 NC	NC time-delay relay contacts. Time delay is after coil is de-energized.
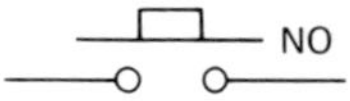 NO	NO momentary contact pushbutton.
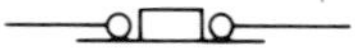 NC	NC momentary open push button.
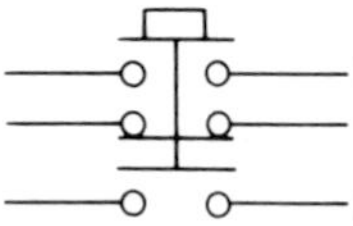 NO NC NO	Momentary contact pushbutton (PB). Two sets of contacts NO, one set NC. Contacts are on same button.
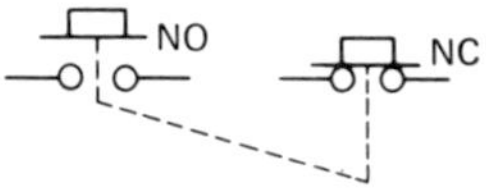 NO NC	Maintained contact pushbutton set. One NO, one NC set of contacts.
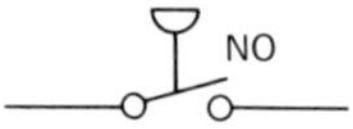 NO	Normally open (NO) pressure switch — make on pressure rise to set point.
 NC	Normally closed (NC) pressure switch — break on pressure rise to set point.
Limit Switches	
 NO	Normally open (NO) limit switch.
 NC	Normally closed (NC) limit switch.
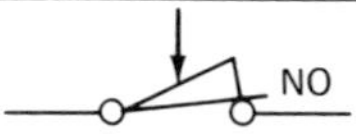 NO	Normally open (NO) limit switch — held closed as when a cam dwells on the limit switch actuator.
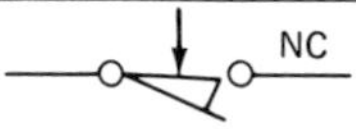 NC	Normally closed (NC) limit switch — held open.

Fig. 9-7. (*Continued*)

are open. The control circuit thus draws no power when the motor is not running. This is an important design consideration.

The relay stick circuit illustrated by rung 1 of the previous example is most useful in designing sequential controls. Use of this circuit gives us a two state memory in that the circuit is either energized or de-energized. Note that the change of state of the circuit is initiated by momentary contact of 1-PB or 2-PB. The pushbuttons could easily have been limit switches and the same

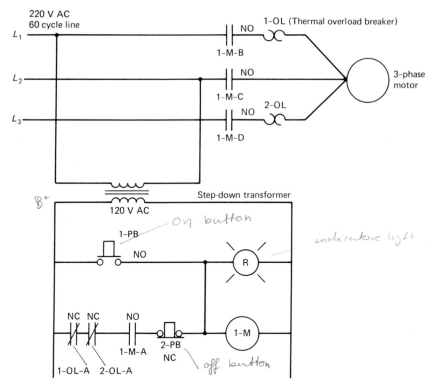

1. When 1-PB is pushed, 1-M coil is energized, pulling in four sets of NO contacts.
2. Current to 1-M coil is maintained through 1-M-A in a stick or holding circuit, even after 1-PB is released.
3. Contacts 1-M-B, 1-M-C, and 1-M-D close, supplying power to each motor phase.
4. If motor overheats, 1-OL or 2-OL break, opening up contacts 1-OL-A or 2-OL-A.
5. To stop motors, push 2-PB, break current to 1-M, allowing it to drop out contacts 1-M-A, 1-M-B, 1-M-C, and 1-M-D, stopping motor. Since 1-M-A opens, circuit will not reactivate when 2-PB is released.

Fig. 9-8. Relay holding circuit.

results would be realized. In general, one control relay (CR) is required for every two-state memory required by the relay sequencing system.

Some common pneumatic cylinder examples are given to illustrate use of relay ladder logic. Example 9-1 shows a double-acting cylinder with a NO limit switch 1-LS located so that the switch closes when the cylinder rod reaches the full extent of its travel. The object of the cylinder control is to cause the cylinder to fully extend, then return home in response to pushing a momentary contact push button, 1-PB.

Recall from Section 6.5.3 that double-acting cylinders are typically controlled by four-way, two-position valves which direct line-pressure air to the appropriate side of the cylinder piston. In this example, the four-way valve is actuated by a double solenoid. The valve spool sticks in its last energized position, and it is important that both solenoids not be energized at the same time. (Refer also to Section 10.1.3 for discussion of limit switches.)

Example 9-2 shows a control circuit which accomplishes the same result, i.e., single-cycle reciprocation of the cylinder in response to pushing 1-PB, but uses a single-solenoid-actuated, spring-return four-way valve. The valve is less expensive, more resistant to accidental spool shifting due to vibration, and will automatically return home in the event of electrical power failure. Also, there is no chance that the operator will burn up the solenoids by holding 1-PB down. The control system is more complex, but the system advantages are probably worth it. In general, cylinder control systems which use double-solenoid-actuated four-way valves are less complex than systems using single solenoid, spring return valves.

Examples 9-3 and 9-4 extend the previous examples to illustrate first the dual-solenoid four-way valve, then the single-solenoid, spring-return four-way valve approach to a *continuously reciprocating* cylinder.

The advantage of the dual-solenoid approach is again only one of control circuit simplicity. When the STOP (2-PB) button is pushed, this cylinder will continue extending or retracting and then remain in either of two positions.

The single-solenoid spring-return valve approach requires a less expensive but more vibration-resistant valve, but it requires an additional control relay and a more complex circuit. When the STOP button is pushed, however, the cylinder will always return to the home position. It is generally desired to have an automatic machine always power down to the same status. This assures correct initial position when powering up, without requiring additional complicated control circuitry.

Note also that two limit switches are required to implement continuous reciprocation. In general, there must be a control signal to initiate each subsequent machine action. For the examples shown here, the signal results from machine position as determined by limit switch closures. The signals to change direction of the cylinder might just as easily have been generated at predetermined time intervals by a timer.

EXAMPLE 9-1
Single-Cycle Reciprocation of a Double-Acting Pneumatic Cylinder

A double-acting pneumatic cylinder is to be used in conjunction with a two-position, four-way directional control valve which is double-solenoid actuated. Upon application of electrical power to one solenoid (solenoid A), the valve switches and stays switched even if power is removed from solenoid A. In order to switch the valve to the other position, power must be applied to solenoid B, and the valve will stay in that position, regardless of power to solenoid B, until switched by solenoid A. Both solenoids must not be energized at the same time, or solenoid coil burn-out will result.

When a momentary-contact pushbutton (1-PB) is pushed, it is desired to have the cylinder extend to a predetermined position (stroke), then return home and stay until the pushbutton is again pushed. The position at which the cylinder is to reverse direction will be sensed by a limit switch, normally open (NO) which will be actuated by a cam on the cylinder rod. The cylinder schematic is as follows:

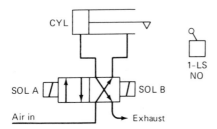

The control circuit is drawn as a relay ladder diagram, with the appropriate control components connected across two power lines. A detailed cylinder control sequence follows the diagram.

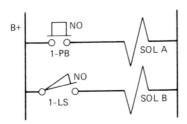

Sequence:

1. Push 1-PB, energizing SOL A, pushing spool to the right so high pressure is directed to the blind end of the cylinder.
2. Air pressure causes the cylinder rod to extend.
3. Release 1-PB, but valve stays in its shifted position. (Note that 1-PB should not be held in longer than the time required for the cylinder to extend, or both SOL A and SOL B would be energized simultaneously).
4. When cylinder cam hits limit switch 1-LS it closes, energizing SOL B and switching valve to the original position, i.e., that shown in the diagram.
5. The piston remains in the home position until 1-PB is pushed again.

EXAMPLE 9-2

Single-Cycle Reciprocation of a Double-Acting Pneumatic Cylinder

It is desired to perform the same cylinder stroke as demonstrated in Example 9-1, but to do so using a single-solenoid-operated, spring-return, two-position four-way valve. Use of a single-solenoid valve avoids the possibility of inadvertently energizing both solenoids at once, thereby damaging the valve. In addition, the cylinder will always return to the home position, even if electrical power is lost. Note that this would not have happened if power were lost during the extension stroke in Example 9-1. The schematic diagram of this system is shown below.

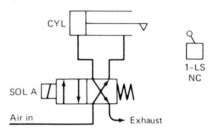

The relay ladder control diagram is shown below. Note that a normally closed (NC) limit switch is used in this control circuit, as well as a control relay (1-CR).

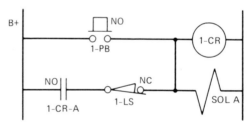

Sequence:

1. Push 1-PB, energizing 1-CR, and pulling in its NO contacts, 1-CR-A.
2. SOL A is energized and stays on by virtue of the 1-CR stick circuit (Fig. 9-8) wherein 1-CR stays on as current flows through its own contact pair, 1-CR-A.
3. Valve is switched to the right, so line pressure extends the cylinder rod.
4. When cylinder rod cam hits NC limit switch 1-LS, the current flowing to relay coil 1-CR is interrupted, causing relay contacts 1-CR-A to drop out or open.
5. With no power to SOL A, the valve spring returns the spool to its original position (that shown in the diagram), and the cylinder returns home.

6. As the cylinder starts to return home, its cam rides off of 1-LS, thereby allowing 1-LS to close. However, 1-CR-A has already opened, so no current flows in the control circuit.

EXAMPLE 9-3

Continuous Reciprocation of a Double-Acting Pneumatic Cylinder Dual Solenoid Control Valve

It is desired to extend the case of Example 9-1 so that the cylinder will continue to extend and retract after pushing a manual START pushbutton. The cycling is to be stopped by pushing a STOP pushbutton. A dual-solenoid directional control valve is to be used as was in Example 9-1.

A control relay, 1-CR, will be used to keep electrical power ON after pushing the START pushbutton, 1-PB. The STOP pushbutton, 2-PB, will cause cycling to cease. Two limit switches will be required, one to signal the control circuit that the cylinder is fully extended (2-LS), and one to signal the circuit that the cylinder is fully retracted (1-LS). Both are normally open (NO), but home position limit switch 1-LS is held closed by the cam when the cylinder is retracted before the start of cycling.

The cylinder schematic diagram below shows the physical location of the limit switches, and defines control valve nomenclature.

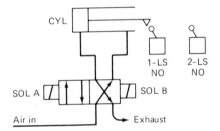

The relay ladder diagram which follows shows how the continuously reciprocating circuit is implemented.

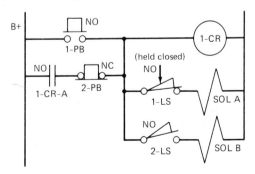

Sequence:

1. Push 1-PB (START) to energize 1-CR, thereby closing 1-CR-A. 1-CR stays on, held through its own contacts, 1-CR-A.
2. SOL A is energized, pushing spool to the right, and applying pressure to the blind end of the cylinder.
3. As cylinder extends, 1-LS opens, de-energizing SOL A, but valve stays in its switched position.
4. When fully extended, the cylinder cam actuates 2-LS, closing it, and applying power to SOL B, and switching valve back to its original position.
5. Valve switches, causing cylinder travel to reverse. Cam rides off 2-LS, but valve stays in its position.
6. When cylinder retracts fully, cam closes 1-LS, thereby energizing SOL A and switching valve again. Cylinder thus extends, and continues to cycle as in steps 3–6 until stopped.
7. When 2-PB (STOP) is pushed, 1-CR is de-energized, so 1-CR-A opens, allowing no power to SOL A or SOL B, even if 1-LS or 2-LS are closed. Cylinder will continue to travel to either the fully extended or fully retracted position, then stop. Note that the cylinder will not always return to the home position to start, so the next start up may have 2-LS held closed to start.

EXAMPLE 9-4

*Continuous Reciprocation of a Double-Acting Pneumatic Cylinder
Single Solenoid Control Valve*

The control system of Example 9-3 may be modified by changing the dual-solenoid directional control valve to a single-solenoid valve. This will result in a control circuit which is slightly more complex (requires another control relay, 2-CR) but which will cause the cylinder to always return home when the STOP (2-PB) is pushed. Note also that 2-LS is now a normally closed (NC) limit switch instead of NO as it was in Example 9-3. The cylinder schematic is:

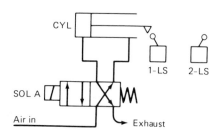

The relay ladder diagram of the control circuit is as follows:

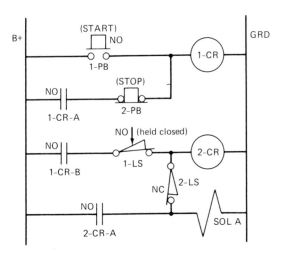

Sequence:

1. Push 1-PB, pulling in 1-CR, and latching through 1-CR-A.
2. 1-CR-B also closes, allowing power to flow to 2-CR, and to SOL A through 2-LS which is NC.
3. 2-CR is also latched closed as its own contacts 2-CR-A close and supply power to 2-CR through 2-LS.
4. SOL A switches the valve spool, so pressure is applied to the cylinder blind end, and it starts to extend.
5. 1-LS, which is NO but held closed in the home position, opens as the cylinder cam extends off of 1-LS, but 2-CR remains on by virtue of 2-CR-A and 2-LS.
6. When the cylinder is fully extended, 2-LS opens, and 2-CR drops out, causing 2-CR-A to open.
7. With 1-LS open and 2-CR-A open, SOL A has no source of power, so the spring switches the valve back to its original position as shown in the diagram. The cylinder starts to retract in consequence.
8. As the cylinder retracts, 2-LS closes, but since 1-LS and 2-CR-A are open, 2-CR and SOL A stay off.
9. When the cylinder cam hits 1-LS, it closes, and applies power to 2-CR which then latches on through 2-CR-A. SOL A gets power, switches the valve, thus the cylinder starts to extend.
10. The cylinder continues to cycle (Steps 5–9) until 2-PB (STOP) is pushed.
11. When 2-PB is pushed, 1-CR drops out, 1-CR-A and 1-CR-B both open. If the cylinder is moving out, 2-CR will be energized, and will

remain so until the cylinder cam hits 2-LS, opening 2-CR and 2-CR-A. The cylinder will return home, but when it hits 1-LS and closes it, it does not switch 2-CR back on because 1-CR-B is open. The cylinder thus stops.

9.2.4. Sequencing Controls with Relay Logic

Examples 9-3 and 9-4 illustrated primitive sequencing circuits in that the cylinder cycled continuously once turned ON. In general, however, sequencing refers to sequential operation of *different* machine elements.

As mentioned early in this chapter, machine operation sequencing can be either time based or event based. There are cases where combinations of the two are appropriate, particularly if one or more of the operations within the machine cycle are time dependent. Use of time-delay relays and/or timers in the relay ladder diagram is not uncommon.

Of principal interest here is event-based sequencing. Previous examples using limit switch closures illustrated initiation of subsequent operations when a machine member (the cylinder rod) reached a particular *position* and closed a limit switch. These were examples of the very common practice of *position sequencing.* Other sensors capable of providing position-based sequence signals include proximity switches and photoelectric switches. A very useful combination device available to the designer is the cylinder with built-in limit switches. This feature is particularly useful if space for standard limit switch placement is limited.

Another event-based method of sequencing is *load sequencing*, wherein the next machine operation is initiated when a machine member exerts a predetermined load (force or torque). When pneumatic or hydraulic cylinders are used, applied cylinder loads can be related to cylinder pressure. Pressure switches (see Section 10.4) are commonly used to provide a sequencing signal when a particular pressure (load) level is reached. Pressure switches are available for which the switching pressure may be adjusted over a wide range, and for which the switching action occurs on either pressure increase to setpoint or decrease to setpoint.

The examples which follow illustrate sequencing of two pneumatic cylinders. The principles can be extended to as many actuators as desired without difficulty. Also, sequencing electric motor or solenoid actions can be performed by substituting these components for the control valve solenoids used in the examples. Keep in mind that most motors are started by energizing the coil of a motor starter, not by applying control-level power directly to the motor. If the motor startup dynamics are unsuitable to the automatic machine cycle, the motor may be run at rated speed, and introduced into the machine drive by means of a clutch. In this case, the control element is most likely a solenoid which directly engages the clutch (and hence the motor) or which shifts a valve to pneumatically or hydraulically engage the clutch. In either case, the relay ladder diagram will be the same.

Example 9-5 considers the two cylinder case following the sequence:

- Initiate cycle with PB;
- Cylinder 1 advance;
- Cylinder 2 advance;
- Cylinder 1 and 2 retract;
- Remain home until PB again pressed.

This sequence implements a typical safety interlock where, for example, cylinder 1 moves a shield or gate into place and cylinder 2 actuates the actual machine operation. After cylinder 2 has fully extended (i.e., performed its operation), it retracts, as does the safety shield.

One of the most often used two-cylinder sequences is called the *clamp-and-work* sequence. The sequence is as follows:

- Advance clamp cylinder until it stalls on workpiece, holding work firmly.
- Advance work cylinder to perform operation on work. Clamp cylinder must remain fully extended.
- Upon completion of work stroke, retract work cylinder. Clamp to remain fully extended.
- When work cylinder clears workpiece, retract clamp cylinder to release work.

Example 9-6 considers the case of one complete clamp-and-work sequence upon actuation by PB. Example 9-7 extends this case to one of continuous cycling of the clamp and work sequence until the STOP PB is pushed. Both examples use dual solenoid 4-way valves to control air pressure to the cylinders. To the greatest extent possible, the "common sense" logic used to develop the ladder diagrams is listed to the right of the ladder. The control sequence is listed below the ladder diagram.

EXAMPLE 9-5
Sequencing Two Cylinders

A safety shield interlock is designed wherein cylinder 1 (CYL 1) moves a safety shield into place. When the shield is in place, cylinder 2 (CYL 2) presses a part onto the workpiece. After the work stroke, both cylinders retract and wait until the operator presses the pushbutton (1-PB) to start the cycle again. Cylinder 1 uses a dual-solenoid control valve, while cylinder 2 uses a single solenoid valve. A NO limit switch is placed at the end of the stroke of both cylinders. The cylinder schematic and relay ladder control circuit follow:

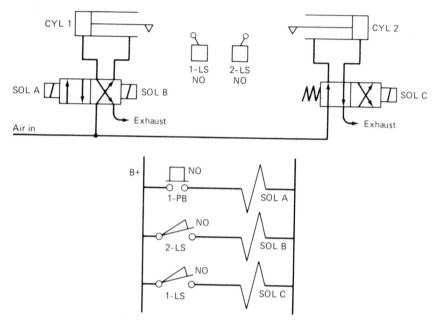

Sequence:

1. Push 1-PB, activate SOL A.
2. CYL 1 extends, actuates 1-LS, energizing SOL C.
3. CYL 2 extends, actuates 2-LS, energizing SOL B.
4. CYL 1 retracts, rides off 1-LS.
5. CYL 2 retracts when 1-LS opens; SOL C is de-energized.
6. Both cylinders dwell at home position until next push of 1-PB.

EXAMPLE 9-6
Clamp-and-Work Sequence
Dual Solenoid Control Valves

This example demonstrates the very common clamp-and-work sequence, wherein the clamp cylinder advances to hold the work, the work cylinder then advances to perform an operation (e.g., drill a hole) on the work, the work cylinder then retracts, and finally, the clamp cylinder retracts, releasing the workpiece. This example uses dual-solenoid valves as shown in the schematic diagram. Note that energizing SOL A advances the clamp cylinder, while activating SOL B retracts it. Energizing SOL C advances the work cylinder, while activating SOL D retracts it. SOL A and SOL B must not be on simultaneously, nor may SOL C be on at the same time as SOL D. Note also that limit switch 1-LS must remain ON while the work cylinder both advances and retracts because the clamp cylinder cam is ·sitting on the switch actuator.

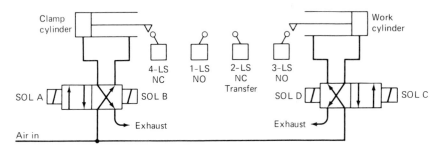

The relay ladder control circuit may be developed as follows:

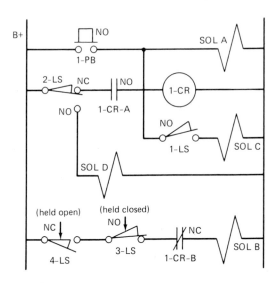

When button is pushed, want SOL A activated, but also need a relay to set up a way to first activate SOL C, then deactivate it while 1-LS stays on.

Transfer contacts break the relay coil circuit before making the SOL D circuit. Upon return to the NC position, 1-CR-A has already opened.

Relay contacts also isolate SOL B during clamp advance, but set it up to be energized by 3-LS on the work cylinder return stroke.

Sequence:

1. Push 1-PB, latch 1-CR through 2-LS and 1-CR-A, and activate SOL A.

2. Clamp cylinder advances, hits 1-LS, activates SOL C.

3. Work cylinder advances, hits 2-LS, dropping out 1-CR, opens 1-CR-A, de-energizes SOL A and SOL C.

4. When 2-LS transfers to NO contact, SOL D is energized, shifting valve, and starting the work cylinder retraction.

5. Work cylinder retracts, closes 3-LS, and energizing SOL B.

6. Clamp valve switches, and clamp cylinder retracts.

7. At home position, 4-LS opens, de-energizing SOL B. Note that no current is flowing in the control circuit in the home position. This is good control circuit design practice.

8. When 1-PB is pushed again, cycle repeats.

EXAMPLE 9-7

Continuous-Cycle Clamp-and-Work Sequence
Dual-Solenoid Control Valves

This example uses the same cylinder schematic as Example 9-6. Instead of returning home after the clamp-and-work cycle, it is desired that the cycle continue until stopped by pushing a STOP PB. The relay ladder control

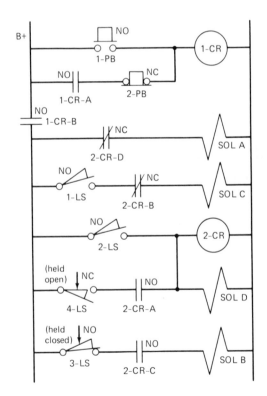

This rung is the stick circuit which turns on the system with 1-PB.

1-CR pulls in 1-CR-B to energize SOL A and start clamp cylinder extension.

When clamp cylinder hits 1-LS, it activates SOL C, thus work cylinder advances.

Work cylinder hits 2-LS, latching in 2-CR through 4-LS, 2-CR-A. SOL A and SOL C are isolated from power as 2-CR-D and 2-CR-B open. 2-CR-C closes, setting up SOL B to activate when work cylinder retracts, hits 3-LS. SOL D is energized, reversing cylinder.

When work cylinder retracts to close 3-LS, SOL B activates, and clamp cylinder retracts, opening 1-LS. When clamp hits 4-LS, 2-CR is dropped out, closing 2-CR-D, energizing SOL A, and starting cycle over again. Cycle continues until 2-PB is pushed.

circuit uses the 1-CR rung solely for the purpose of turning the cycle on and off. Notice that this rung serves the same purpose as a maintained contact switch which supplies power to the control ladder.

Figure 9-9 presents a control sequence diagram which shows the state of each control element as a function of time through one cycle of Example 9-6. This diagram allows visualization of which devices are energized at what times, and helps assure that opposing valve solenoids are not energized simultaneously.

Example 9-8 illustrates a simplification of the clamp-and-work control sequence of Example 9-6 through use of *impulse limit switches*. These limit switches produce a pulse of current sufficient to shift the solenoid-operated spool valve, but do not maintain the current, even though the cylinder cam might dwell on the limit switch actuator. Impulse limit switches are discussed in Section 10.2.

A final example of two-cylinder sequencing is presented in Example 9-9. Again, the clamp-and-work sequence is illustrated, but it is mechanized using

Control Device	Clamp Cylinder Extend	Work Cylinder Extend	Work Cylinder Retract	Clamp Cylinder Retract
1. Push 1-PB to START	▯			
2. SOL A ON				
3. 1-CR ON				
4. 1-CR-B (NC)			Contacts closed	
5. 1-CR-A (NO)	Contacts closed			
6. 1-LS (NO) Closed				
7. SOL C ON				
8. 2-LS (NC) Closed				
9. 2-LS (NO) Closed		▯		
10. SOL D ON		▯		
11. 3-LS (NO) Closed				
12. SOL B ON				
13. 4-LS (NC) Closed				

Fig. 9-9. Sequence diagram of a clamp-and-work cycle.

single-solenoid, spring-return control valves. The relay ladder diagram is developed using the common-sense trial-and-error approach which the designer would follow in designing the control system.

EXAMPLE 9-8

Circuit Simplification with Impulse Switches

The relay ladder control circuit described in Example 9-6 may be simplified using *impulse limit switches* (see Section 10.1.3). The normally open (NO) impulse switch closes its contacts momentarily, then opens them again, regardless of how long the moving cam may dwell on the switch actuator. This type of switch works well with dual-solenoid valves, since these valves need only a short energization of one or the other solenoid to transfer the valve spool. The cylinder circuit schematic and relay ladder diagram follow:

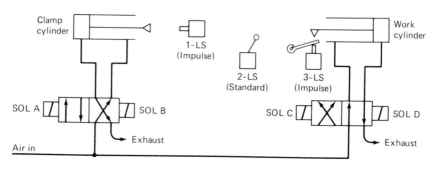

Sequence:

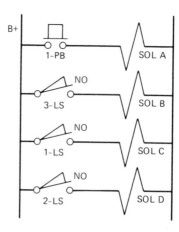

1. Push 1-PB, energize SOL A, extend clamp.
2. Clamp hits 1-LS (impulse) switches SOL C, so work cylinder extends.
3. Work cylinder hits 2-LS, energizing SOL D and reversing its direction.
4. Work cylinder returns home hits 3-LS, energizes SOL B, and clamp retracts.
5. Cycle is restarted by 1-PB even though work cylinder sits on 3-LS. The impulse switch de-energizes SOL B after making momentary contact to shift clamp valve.

EXAMPLE 9-9
Clamp-and-Work Sequence with Single-Solenoid Control Valves
   A Comprehensive Example

In general, it is good design practice to use single-solenoid directional control valves since they cause the cylinders to return home automatically in the event of an electrical power failure. This example demonstrates a control circuit design for the now familiar clamp-and-work sequence using single-solenoid, spring-return valves. The control circuit is developed using the common-sense method, and the reasoning for each step is discussed.

Assume that four (4) limit switches will be required, and draw the cylinder schematic as follows:

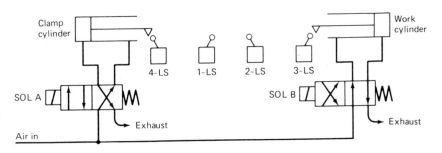

Develop the control circuit as follows:

RUNG 1. When the START PB (1-PB) is pushed, use a 1-CR NO contact to latch the circuit on. Also, there is no reason why the 1-PB should not also start the cycle by actuating SOL A. Note that SOL A must be ON for both the clamp and work strokes, and that it must be turned OFF in order for the clamp cylinder to retract. This implies that there should be a set of contacts in the circuit which are closed to start, but which open when the work cylinder retracts. We can therefore draw the first rung of the ladder as follows, with the NC contacts yet to be identified:

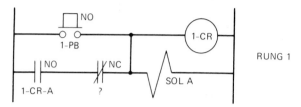

RUNG 2. When the clamp cylinder extends, we want the signal from 1-LS to activate SOL B so that the work cylinder will extend. In order to retract the work cylinder, SOL B must be de-energized. Since 1-LS will be actuated by the clamp cylinder for *both* work cylinder strokes, a NC set of contacts must be placed in the circuit to open at the appropriate time and de-energize SOL B. Again, these contacts will be identified later.

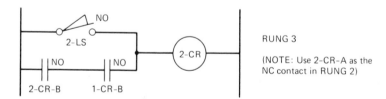

RUNG 3. We now want the work cylinder to signal its own reversal when it extends to hit 2-LS. This could be done by simply closing a NO 2-LS to activate a relay (2-CR) which controls the NC contacts shown in RUNG 2. However, as the work cylinder retracts, opening 2-LS, the relay coil 2-CR will open, its NC contacts will close, and SOL B will again be activated. Therefore, another set of 2-CR contacts which are NO should be used to latch 2-CR ON. One more set of contacts are required—a set to release 2-CR at the proper time, i.e., after the clamp cylinder has retracted off 1-LS. This could be accomplished by using a set of 1-CR NO contacts which would be closed at the time 2-CR was actuated, and would deactivate 2-CR when 1-CR was turned OFF to retract the clamp cylinder. Therefore, RUNG 3 would look like:

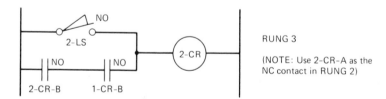

RUNG 4. We now need to design a way for SOL A to be released when the work cylinder returns home, actuating 3-LS. This could be done by

having 3-LS actuate a third relay coil 3-CR, and using this coil's NC contacts back in RUNG 1 to drop out 1-CR, thereby releasing SOL A. Note, however, that the work cylinder sits on 3-LS at the start of the cycle, and would never allow the clamp cylinder to extend unless another set of contacts are put into RUNG 4 which are open until after the work cylinder leaves 3-LS, but which close before the work cylinder returns to 3-LS. A set of 2-CR contacts would be perfect, since 2-CR is activated when the work cylinder extends, hitting 2-LS. Thus RUNG 4 would be:

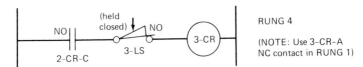

RUNG 4

(NOTE: Use 3-CR-A
NC contact in RUNG 1)

Putting all of the rungs together, the complete ladder diagram would be:

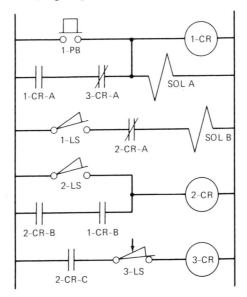

Note:

1. The system will go through one cycle, return home, and wait for another 1-PB command to cycle again.
2. For single-cycle operation, *the fourth limit switch 4-LS is not necessary.*

It is now desired to make the circuit cycle continuously until turned OFF by a second momentary-contact pushbutton, 2-PB. The first method which comes to mind would be to place the fourth limit switch, 4-LS in parallel with 1-PB, so that every time the clamp cylinder returned home, it would "make" 4-LS, thus start the cycle over. However, a momentary open STOP pushbutton could not stop the cycle. At the home position, 4-LS would always start the cycle up again. Therefore, a set of NO contacts need to be placed in series with 4-LS which, when open, stop the cycle. These contacts need to be closed after the cycle starts up initially, and remain ON until the STOP pushbutton (2-PB) is pushed. This could be accomplished with a fourth relay, 4-CR, in a latching circuit which is latched after the cycle starts, say by 2-CR or 3-CR, and which can be broken by 2-PB. If RUNG 1 of the previous diagram is modified, and RUNG 5 added, the following circuit results for continuous reciprocation:

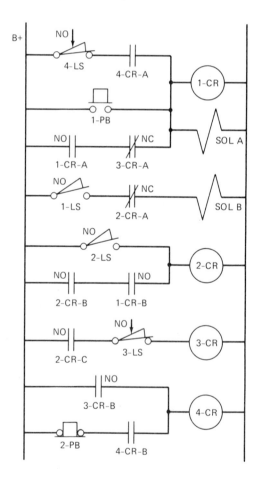

9.3. SWITCHING LOGIC

The previous section showed how machine sequence control systems could be designed using the relay ladder diagram technique. The approach to ladder diagram design was a combination of common sense with trial and error. It was relatively easy to design a system which would work, but one quickly learns that there are always several ways to design a control system to follow the same sequence, even using the same control components. This leads one to ask if there is indeed an optimum control system design for a specific machine sequencing problem.

Notice that all machine sequencing control systems discussed so far use control components which are either ON or OFF. As such, these control systems could be modelled and analyzed quite conveniently by a mathematical system which is concerned with only two variables, i.e., ON and OFF.

Boolean algebra or *switching algebra* is the mathematics of variables which have only two states: ON or OFF, HIGH or LOW, 1 or 0, TRUE or FALSE. It turns out that we can express the same control relationships mathematically using Boolean equations as we did schematically using relay ladder diagrams. Furthermore, there exist orderly control system design techniques which replace the common-sense/trial-and-error approach and result in the most basic control system design possible to achieve the required control. This is not to say that the common-sense approach does not lead to the "correct" control system—it does. However, in very complex control problems, the common-sense approach may achieve the desired result in a redundant manner—i.e., using more control relays or limit switches than are really necessary.

9.3.1. Boolean Algebra

A *function* in ordinary algebra is a statement of the value of a *dependent* variable based on the values of one or more *independent* variables as related by the function. In ordinary algebra, the function $y = 3x + z$ yields an infinite number of numerical values for y (the dependent variable), based on the infinite number of choices for x and y (the independent variables).

The dependent variable of a Boolean algebra function can have only one of two values—1 or 0—depending on the function itself and the value of the independent variables. The independent variables themselves can have only one of two values also, namely, 1 or 0. In this sense, Boolean algebra is much simpler than ordinary algebra.

Keep in mind that a mathematical function is not reality—it is an abstract representation of reality. What is real in machine control systems are the control components (relays, limit switches, solenoids, etc.) and the manner in which they are interconnected. The independent variables in a Boolean representation of a system will be the *control inputs,* which include limit switches, relay contacts, or manual push buttons. These devices will be either ON or OFF depending on the state of the machine. The dependent variables will be *control outputs,* such as solenoids, relay coils, lights, motors, etc. Their state will be either ON or OFF, depending on the state of the control inputs and the manner in which the inputs are connected. The manner in which control inputs and control outputs are connected defines the *control function.*

There are four basic Boolean algebra functions from which all others can be derived. They are:

1. YES function (also called *equals* or *is*)
2. NOT function (also called *inverse* or *complement*)
3. AND function (also called *logical product*)
4. OR function (also called *logical sum*)

Figure 9-10 describes each function in terms of control element inputs and outputs. Fig. 9-10 also describes two auxiliary control functions which, in conjunction with the four basic functions, are *sufficient* to solve *all* digital automation control problems. These are the MEMORY function, which is realized electronically by a flip-flop; and TIMING, which is mechanized by a time-delay relay, a timer, or a digital counter.

There are two ways of expressing the functional relationship between switching variables: the Boolean algebra equation, and the *truth table.* Figs. 9-11 through 9-14 show the logic symbols associated with each of the four basic Boolean functions, the truth table associated with each function, and the relay ladder diagram equivalent of each function. Note that the relay ladder control inputs are considered OFF when they are not actuated, and ON when they are actuated. Thus a NC limit switch such as that in Figure 9-12 is OFF in a logical sense even though it is conducting or ON in a real sense.

Two very common logic functions are the NAND and NOR functions which are derived by compounding the NOT and AND functions and the NOT and OR functions respectively. Figs. 9-15 and 9-16 show the logic symbols and truth tables for these functions. Notice that the truth table for the NAND function is the opposite or *complement* of that for the AND function (Fig. 9-13) and that the NOR function is indeed the complement of the OR function (Fig. 9-14).

Boolean algebra has three basic laws (commutative, associative, and distributive) which dictate the ways in which Boolean variables may be

Basic Control Element Functions

Function	Symbol	Element Operation
YES	Control input → YES → Control output	Control output is ON if control input is ON. Output is OFF when input is OFF.
NOT	Control input → NOT → Control output	Control output is ON when input is OFF, output is OFF when input is ON.
AND	Input A → Input B → Input C → AND → Control output ⋮ Input N →	Control output is ON when all control inputs (which may range from 2 to N) are ON; i.e., output is ON when A AND B AND C are ON.
OR	Input A → Input B → Input C → OR → Control output ⋮ Input N →	Control output is ON if any one OR more of the inputs are ON.

Auxiliary Control Element Functions

Function	Symbol	Element Operation
MEMORY (Flip-flop)	Input A → MEMORY → Control output Input B →	Control output is ON if input A was ON last. Output is OFF if input B was ON last.
TIMING	Control input → TIMING (delay) → Control output	Control output is ON after a time delay following control input ON.

Fig. 9-10. Logical functions mechanized by control devices.

YES function represents logical equality

Symbol

Fig. 9-11. The YES function

A	A
Input	Output
0	0
1	1

Truth Table

The truth table shows in graphical form all possible inputs and their resulting outputs.

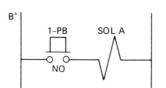

Relay Ladder Analogy

SOL A (the control output) is OFF (or 0) when 1-PB (the control input) is OFF.

SOL A is ON when 1-PB is ON.

Therefore, the Boolean equation relating 1-PB and SOL A for this ladder rung is:

$$\text{SOL A} = \text{1-PB}$$

Fig. 9-11. (*Continued*)

The NOT function represents logical inverse

Symbol

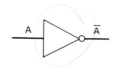

The bar over A represents the *inverse* or *complement* of A.

The small circle at the tip of the NOT symbol differentiates it from the YES symbol.

Truth Table

A	$\overline{A}$
Input	Output
0	1
1	0

Note that the output is always the inverse or complement of the input.

Relay Ladder Analogy

SOL A (control output) is ON when I-LS (control input) is OFF, i.e., not actuated.

SOL A is OFF when I-LS is ON, i.e., actuated.

The Boolean equation for this ladder rung is:

$$\text{SOL A} = \overline{\text{I-LS}}$$

(SOL A equals NOT I-LS).

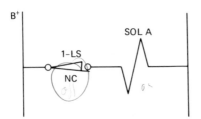

Fig. 9-12. The NOT function.

AND Function represents the logical product

Symbol

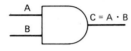

Output C is ON if A AND B are ON. An unlimited number of inputs are possible, shown symbolically by additional input lines on the symbol. Only one output is possible, and it is ON only if *all* of the inputs are true.

Inputs		Output
A	B	C
0	0	0
0	1	0
1	0	0
1	1	1

Truth Table

Relay Ladder Analogy

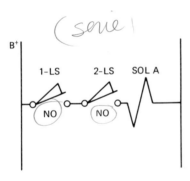

SOL A is ON only when 1-LS AND 2-LS are both on (i.e., actuated). SOL A is OFF for any other combination of control inputs.

The Boolean equation for this ladder rung is:

$$\text{SOL A} = \text{1-LS} \cdot \text{2-LS}$$

(SOL A equals 1-LS AND 2-LS).

In general, two or more NO switches or other control inputs in series create the AND analogy.

Fig. 9-13. The AND function.

OR Function represents the logical sum

Symbol

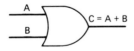

Output C is ON if input A OR B (or both) is ON. Symbol may show unlimited inputs, but only one output. Output is ON if any one (or more) of the inputs are ON.

Fig. 9-14. The OR function.

Inputs		Output
A	B	C
0	0	0
0	1	1
1	0	1
1	1	1

Truth Table

(parallel)

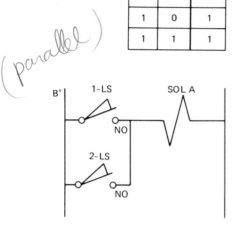

Relay Ladder Analogy

SOL A is ON if 1-LS OR 2-LS is ON, or if both are ON (i.e. actuated).

The Boolean equation is:

$$SOL\ A = 1\text{-}LS + 2\text{-}LS$$

(SOL A equals 1-LS OR 2-LS).

In general, two or more NO switches or other control inputs in parallel create the OR analogy.

Fig. 9-14. (*Continued*)

The NAND Function represents the combination of the AND and NOT functions:

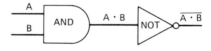

Symbol

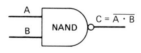

$$C = \overline{A \cdot B}$$

Inputs		Output
A	B	C
0	0	1
0	1	1
1	0	1
1	1	0

Truth Table

Fig. 9-15. The NAND function.

Relay Ladder Analogy

SOL A is ON when 1-LS is OFF (i.e., not actuated) AND 2-LS is OFF (i.e., not actuated).

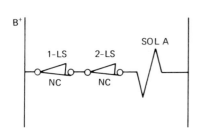

Boolean equation for this ladder rung is:

$$\text{SOL A} = \overline{1\text{-LS}} \cdot \overline{2\text{-LS}} = \overline{1\text{-LS} \cdot 2\text{-LS}}$$

(SOL A equals 1-LS NAND 2-LS).

In general, two or more NC switches or other control inputs in series create the NAND analogy.

Fig. 9-15. (*Continued*)

NOR Function represents the combination of the OR and NOT functions:

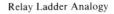

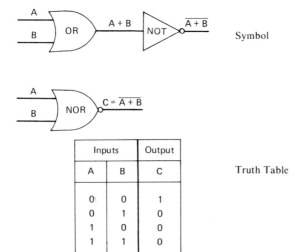

Symbol

Truth Table

Relay Ladder Analogy

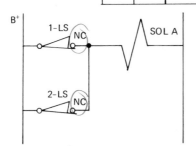

SOL A is ON when 1-LS is OFF (i.e., not actuated) OR 2-LS is OFF (i.e., not actuated).

Boolean equation for this rung is:

$$\text{SOL A} = \overline{1\text{-LS} + 2\text{-LS}} = \overline{1\text{-LS}} + \overline{2\text{-LS}}$$

(SOL A equals 1-LS NOR 2-LS).

Fig. 9-16. The NOR function.

TABLE 9-3.   Boolean Algebra Identities.

Basic Boolean Algebra Laws

1. $A + B = B + A$ } Commutative laws; i.e., it does not matter in which order variables are
    $A \cdot B = B \cdot A$ } written.
2. $A + B + C = A + (B + C) = (A + B) + C$ } Associative laws, i.e., it does not matter
    $A \cdot B \cdot C = A \cdot (B \cdot C) = (A \cdot B) \cdot C$ } how variables are grouped.
3. $A \cdot (B + C) = (A \cdot B) + (A \cdot C)$ } Distributive laws; i.e., binary variables can be fac-
    $A + (B \cdot C) = (A + B) \cdot (A + C)$ } tored somewhat like ordinary algebra variables.

Boolean Algebra Identities

1. $A + A = A$
2. $A \cdot A = A$
3. $A \cdot \bar{A} = 0$
4. $A + 1 = 1$
5. $A \cdot 1 = A$
6. $A + 0 = A$
7. $A \cdot 0 = 0$
8. $A + \bar{A} = 1$
9. $\bar{\bar{A}} = A$
10. $A + AB = A$ $A (B + 1) =$
11. $AB + A\bar{B} = A$
12. $A + B = \overline{\bar{A} \cdot \bar{B}}$ } DeMorgan's Theorem
13. $AB = \overline{\bar{A} + \bar{B}}$ }
14. $A(\bar{A} + B) = AB$
15. $A + \bar{A}B = A + B$
16. $\bar{A} + AB = \bar{A} + B$
17. $\bar{A} + A\bar{B} = \bar{A} + \bar{B}$
18. $(A + B)(A + \bar{B}) = A$
19. $AC + AB + B\bar{C} = AC + B\bar{C}$
20. $(A + B) \cdot (B + C) \cdot (\bar{A} + C) = (A + B) \cdot (\bar{A} + C)$

manipulated. These are shown in Table 9-3. Also shown in Table 9-3 are twenty Boolean identities which are useful in manipulating or simplifying complex Boolean expressions.

The practical use of the identities is one of control-component elimination through simplification of the overall control equation or control *algorithm*. If the Boolean equations are derived from either (a) the relay ladder diagram or (b) a truth table, they *may* be simplified, then used to reconstruct the actual control system using fewer (nonredundant) components.

9.3.2.   Karnaugh Mapping

The Karnaugh map is a graphical representation of a truth table which is often helpful in visualizing logic functions. It consists of a matrix of all possible inputs including each of their possible states (i.e., ON or OFF).

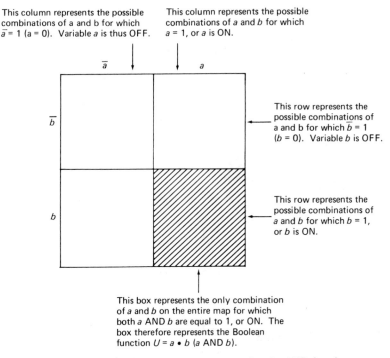

This column represents the possible combinations of a and b for which $\bar{a} = 1$ (a = 0). Variable a is thus OFF.

This column represents the possible combinations of a and b for which a = 1, or a is ON.

This row represents the possible combinations of a and b for which $\bar{b} = 1$ (b = 0). Variable b is OFF.

This row represents the possible combinations of a and b for which b = 1, or b is ON.

This box represents the only combination of a and b on the entire map for which both a AND b are equal to 1, or ON. The box therefore represents the Boolean function $U = a \cdot b$ (a AND b).

Fig. 9-17. Two-variable Karnaugh map representing the AND function.

Figure 9-17 shows a two-variable (a,b) Karnaugh map. If we denote the boxes in the matrix where $a = 1$ and $b = 1$ by cross-hatching, then we see that only one box represents the state wherein *both a* AND *b* are equal to 1. This box then represents the logical product, $a \cdot b = 1$.

Figure 9-18 is also a two-variable Karnaugh map showing the function a OR $b = 1$. Note that this function is represented by three boxes in the map, where either a or b or both a and b are equal to 1.

Figure 9-19 shows the extension of Karnaugh maps to 3 variables, a, b, c. Notice that the two-variable map is "folded" out about an axis of symmetry. The extension to four variables simply requires more "openings" of the map. Notice that there will be 2^N boxes, where N is the number of input variables under consideration. Examples 9-10 and 9-11 show the representation of three- and four-variable Boolean functions on Karnaugh maps. It is important to note that any given combination of control inputs can be represented by one box in the Karnaugh map.

Example 9-12 provides a comprehensive Karnaugh map analysis of the clamp-and-work cylinder sequencing problem of Example 9-7. The proce-

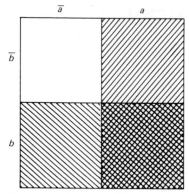

The right-hand (shaded) column represents the combinations of a and b where $a = 1$ (a is ON), and the lower row represents both possibilities for $b = 1$ (b is ON). Therefore, the total shaded area (three boxes) represents the possibilities where a OR b are

The total shaded area represents the Boolean function

$$U = a + b \quad (a \text{ OR } b)$$

Fig. 9-18. Two-variable Karnaugh map representing the OR function.

dure described shows how the Karnaugh map can be used to visualize the control input device states at every point in the machine cycle. In addition, the Karnaugh map technique allows one to easily write the Boolean algebra equations relating control outputs to the state of control inputs. Finally, the example shows how the Boolean equations are used to construct the relay

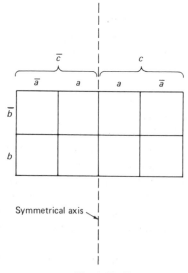

The two-variable (a,b) Karnaugh map can be extended to three variables (a,b,c) by "opening up" the map to create a mirror image about a symmetrical axis. The original two-variable map then represents one state of the new variable (i.e., $\bar{c}$), and the "unfolded" map becomes the representation of c.

Additional variables are considered by simply continuing to "unfold" the map.

Fig. 9-19. Three-variable Karnaugh map.

ladder diagram which, indeed, is the same as that derived using other design methods.

EXAMPLE 9-10
Three Variable Karnaugh Map Representation of a Boolean Function

Represent the Boolean expression $U = a + b + \bar{c}$ (a OR b OR NOT c) using a three-variable Karnaugh map.
Represent the three variables by shading boxes as follows:

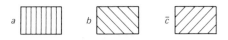

Draw a three-variable map, and shade all areas appropriately:

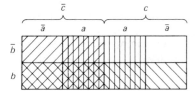

The function $U = a + b + \bar{c}$ is represented by all the shaded areas together. Notice that the same function could also be represented by the area which is *not* shaded as follows:

$$U = \overline{\text{(area not shaded)}} \qquad \text{(i.e., the complement of the non-shaded area)}$$
$$= \overline{\bar{a} \cdot \bar{b} \cdot c}$$
$$= a \cdot b \cdot \bar{c}$$

EXAMPLE 9-11
Four-Variable Karnaugh Map Representation of a Boolean Function

Represent the Boolean function $U = a \cdot b \cdot c \cdot d$ (a AND b AND c AND d) using a four-variable Karnaugh map.
The logical product of several variables is the *intersection* of the shaded area representing each of the variables. If each variable is represented by shaded area as follows:

then the following Karnaugh map results:

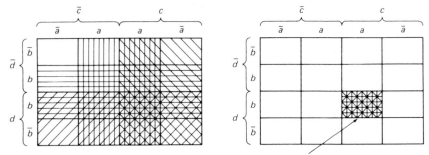

The intersection of the shaded areas represents the function:

$$U = a \cdot b \cdot c \cdot d$$

(other shaded areas have been removed for clarity).

EXAMPLE 9-12

Use of Karnaugh Maps in Control Cycle Analysis

Consider the two cylinder clamp and work sequence illustrated in Example 9-7. The cylinder schematic diagram is reproduced here for convenience:

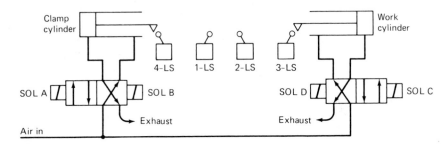

To simplify the problem somewhat, the START pushbutton (1-PB) will be neglected. As a practical matter, we can always start a control circuit with a PB, and if necessary, a CR stick circuit.

There are then four control inputs to the circuit: 1-LS, 2-LS, 3-LS, and 4-LS. There are four control outputs: SOL A, SOL B, SOL C, SOL D. A four-variable Karnaugh map is then drawn, using the LS control inputs as variables. Without consideration for whether the LS is NO or NC, LS represents the *actuated* state, and $\overline{LS}$ represents the unactuated state of the limit switch. It is convenient to represent both 1-LS and 4-LS as columns and 2-LS and 3-LS as rows, because neither of these pairs can physically be actuated at the same time, i.e., 1-LS and 4-LS cannot both be actuated at the same time because the actuating cam cannot be in two different positions at the same time. The same is true for 2-LS and 3-LS.

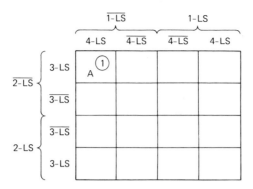

Each cell in the Karnaugh map represents a particular combination of control inputs, or a *control state*. By the nature of the Karnaugh Map, each possible combination has it own cell, so that we could represent that state uniquely by that cell. At the start of the cycle, 4-LS and 3-LS will be actuated, or ON, because the cylinders are at their home positions. Point 1 on the preceding map represents this state. Also, since this is the cycle start point, we want SOL A to be energized, so A is written in the same cell to designate the required control output for this particular combination of control inputs.

If we then continue to plot control input states as the cycle progresses, we construct a cycle path on the map:

Point	Control Input State	Cylinder Condition
2	$\overline{4\text{-LS}}$ $\overline{1\text{-LS}}$ $\overline{2\text{-LS}}$ 3-LS	Clamp cylinder extending
3	$\overline{4\text{-LS}}$ 1-LS $\overline{2\text{-LS}}$ 3-LS	Clamp hits 1-LS
4	$\overline{4\text{-LS}}$ 1-LS $\overline{2\text{-LS}}$ $\overline{3\text{-LS}}$	Work cylinder extending
5	$\overline{4\text{-LS}}$ 1-LS 2-LS $\overline{3\text{-LS}}$	Work hits 2-LS
6	$\overline{4\text{-LS}}$ 1-LS $\overline{2\text{-LS}}$ $\overline{3\text{-LS}}$	Work cylinder retracting
7	$\overline{4\text{-LS}}$ 1-LS 2-LS $\overline{3\text{-LS}}$	Work hits 3-LS
8	$\overline{4\text{-LS}}$ $\overline{1\text{-LS}}$ $\overline{2\text{-LS}}$ 3-LS	Clamp cylinder retracting
9	4-LS $\overline{1\text{-LS}}$ $\overline{2\text{-LS}}$ 3-LS	Clamp hits 4-LS

Notice that the condition at points 3 and 7 (i.e., the same cell) is one of requiring two separate outputs for the same combination of control inputs. We want SOL C to energize if the clamp cylinder has just extended. We want SOL B to energize if the work cylinder has just retracted.

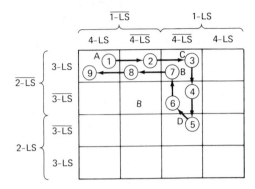

Nevertheless, we have 1-LS and 3-LS actuated at the same time regardless of which output we desire. The control system literally does not know if the cylinders are coming or going without additional information. What is required is an additional control variable which distinguishes between point 3 and point 7 and allows a single control output (the correct one) to be made for an unambiguous set of control inputs. This additional control variable is called *memory*, and will be represented by M for *memory actuated*, and $\overline{M}$ for *Memory non-actuated*. The control output M_1 represents the signal which switches memory from $\overline{M}$ to M (i.e., from OFF to ON), and M_2 represents the signal which switches memory back from M to $\overline{M}$.

We therefore expand the four-variable Karnaugh map to a five-variable map, and map out the control cycle as shown below. Note that Karnaugh maps should be thought of as continuous from top to bottom and from edge to edge, as though the map were plotted on a sphere. Thus position 9 is really adjacent to position 1.

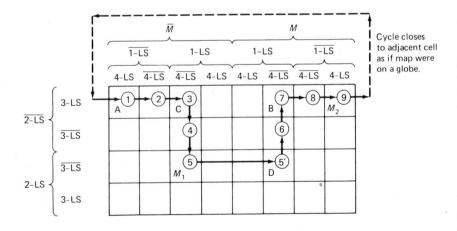

Cycle closes to adjacent cell as if map were on a globe.

By inspection of the map, we can then write the Boolean expressions for control outputs as a function of control inputs:

$$\text{SOL A} = 3\text{-LS} \cdot 4\text{-LS} \cdot \overline{1\text{-LS}} \cdot \overline{2\text{-LS}} \cdot \overline{M}$$
$$\text{SOL C} = 3\text{-LS} \cdot 1\text{-LS} \cdot \overline{4\text{-LS}} \cdot \overline{2\text{-LS}} \cdot \overline{M}$$
$$M_1 = 1\text{-LS} \cdot 2\text{-LS} \cdot \overline{3\text{-LS}} \cdot \overline{4\text{-LS}} \cdot \overline{M}$$
$$\text{SOL D} = 1\text{-LS} \cdot 2\text{-LS} \cdot \overline{3\text{-LS}} \cdot \overline{4\text{-LS}} \cdot M$$
$$\text{SOL B} = 1\text{-LS} \cdot 3\text{-LS} \cdot \overline{2\text{-LS}} \cdot \overline{4\text{-LS}} \cdot M$$
$$M_2 = 3\text{-LS} \cdot 4\text{-LS} \cdot \overline{1\text{-LS}} \cdot \overline{2\text{-LS}} \cdot M.$$

These equations are not simplified, and contain redundancies in many cases. For example, 1-LS ON means 4-LS must be OFF, and vice versa. There is no need to specify 1-LS $\cdot$ $\overline{4\text{-LS}}$ (i.e., 1-LS AND NOT 4-LS) since the same input will be realized if we specify only 1-LS.

The first level of simplification can take place by inspection of the Karnaugh map, realizing that we need only specify the simplest combination of possible control inputs into which our desired output uniquely falls. For example: SOL A energization (point 1) can be specified as:

$$\text{SOL A} = 4\text{-LS} \cdot \overline{M}.$$

This says that we will energize SOL A for any combinations of input variables which fall into column 1 and column 4 of the previous Karnaugh map. Notice, however, that the only combination of input control variables which can occur in those two columns is that of cell 1. All other possible combinations in those two columns are prevented from occurring by the setup of the control system (i.e., the placement of limit switches). The fact is that we do not care if those other states can cause SOL A to turn ON, simply because those combinations of inputs (states) cannot occur. These are commonly referred to as "don't care" states in control system design.

Continuing the simplification of the above six Boolean equations by inspection of the map yields:

$$\text{SOL C} = 1\text{-LS} \cdot \overline{M} \cdot 3\text{-LS}$$
$$M_1 = 2\text{-LS} \cdot \overline{M}$$
$$\text{SOL D} = 2\text{-LS} \cdot M$$
$$\text{SOL B} = 1\text{-LS} \cdot 3\text{-LS} \cdot M$$
$$M_2 = 4\text{-LS} \cdot M.$$

A second round of simplifications may be made by considering the practical aspects of the control system, and questioning the need for input states to singularly define the desired outputs.

1. When switching from state M to $\overline{M}$ (9 to 1), SOL A must be energized to start the cycle over again, therefore the memory switching signal alone should be sufficient to start SOL A. Thus:

$$SOL\ A = \overline{M}$$

2. Since 3-LS is at home when 1-LS is actuated and we want SOL C energized, 3-LS is not necessary as a control input to energize SOL C. Thus:

$$SOL\ C = 1\text{-}LS \cdot \overline{M}.$$

3. The only input signal necessary to shift the memory $\overline{M}$ ON to M is actuation of 2-LS. The term 2-LS $\cdot$ $\overline{M}$ is redundant, so that:

$$M_1 = 2\text{-}LS.$$

4. SOL D could be energized as a result of 2-LS actuation, but since 2-LS has just switched M ON, specification of both variables is not necessary. All that is required is:

$$SOL\ D = M.$$

5. SOL B must energize when the work cylinder returns home and actuates 3-LS. The fact that 3-LS is ON *and* the memory, M is ON is sufficient to uniquely define the signal, thus:

$$SOL\ B = 3\text{-}LS \cdot M.$$

6. Finally, to return to position 1, we must shift memory from ON (M) to OFF ($\overline{M}$). The only signal required to do that is that of 4-LS, since we always want $\overline{M}$ (i.e., memory OFF) when the clamp cylinder is home and 4-LS is actuated. Therefore:

$$M_2 = 4\text{-}LS.$$

The simplified set of Boolean equations defining the control system are therefore:

Eqn.	Control Output	Control Input	Activity
1	SOL A $= \overline{M}$		Starts clamp out
2	SOL C $= $ 1LS $\cdot \overline{M}$		Starts work cylinder out
3	$M_1 = $ 2-LS		Switches $\overline{M}$ to M
4	SOL D $= M$		Starts work back
5	SOL B $= $ 3-LS $\cdot M$		Starts clamp back
6	$M_2 = $ 4-LS		Switches M to $\overline{M}$

If we use 2-CR as a mechanization of the memory, and define the state M as that when 2-CR is ON, $\overline{M}$ as that when 2-CR is OFF, we can draw a relay ladder diagram to implement these Boolean equations. Recall that we neglected 1-PB for starting the circuit as well as 2-PB for stopping the cycle. Therefore, we will arbitrarily start the ladder diagram with 1-PB and 1-CR.

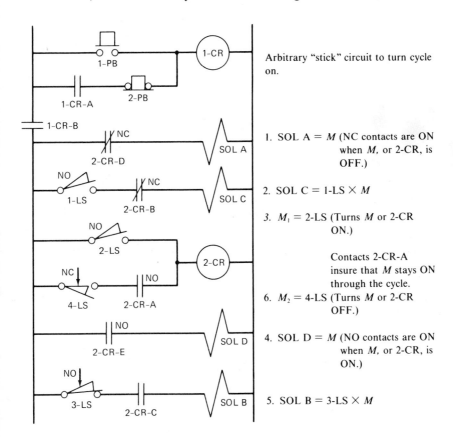

Arbitrary "stick" circuit to turn cycle on.

1. SOL A $= M$ (NC contacts are ON when M, or 2-CR, is OFF.)

2. SOL C $= $ 1-LS $\times M$

3. $M_1 = $ 2-LS (Turns M or 2-CR ON.)

Contacts 2-CR-A insure that M stays ON through the cycle.

6. $M_2 = $ 4-LS (Turns M or 2-CR OFF.)

4. SOL D $= M$ (NO contacts are ON when M, or 2-CR, is ON.)

5. SOL B $= $ 3-LS $\times M$

Note that SOL D is to turn on when 2-CR turns ON. If the ladder rung containing SOL D were combined with the rung containing 2-CR, the resultant ladder diagram would be identical to that derived for Example 9-7 using common-sense reasoning.

9.4. ELECTRONIC LOGIC

After the designer arrives at a set of Boolean control equations, whether derived from the relay ladder diagram, from a truth table, or from Karnaugh mapping techniques, he must then *mechanize* the logic defined by those equations. Mechanization means selection of control components (electrome-chanical, electronic, or otherwise) which yield the desired control outputs for the given control inputs.

Relay logic was mechanized with components (relays, contactors, limit switches, etc.) which were capable of carrying enough power to switch actuator level power directly. In fact, the relays themselves serve as amplifiers whereby high output power (voltage and currents) is switched with low input- (or control)-level power. Note however, that relay logic systems require time (10–100 ms per relay) to switch, and the relays themselves are not highly reliable.

The next logical advance in machine control technology was to use solid state devices (with their high reliability, fast switching, and lower control power levels) to mechanize machine logic. A very wide selection of solid state devices are available, usually packaged as Dual In-line Packages (DIPs), which perform the basic Boolean algebra functions as well as many variations, combinations, and convenient additions to same. Electronic logic devices are called *gates* because the output terminal exhibits a voltage (and passes current) only when the inputs to the gate are true.

Logic gates operate at power levels which are compatible with the solid state devices which mechanize the logic. The most popular discrete compo-nent logic family is TTL (Transistor–Transistor Logic), which interprets the logical 1 or ON condition to be 2.4–5.0 volts DC, and the logical 0 or OFF condition to be 0–0.4 volts DC. TTL devices must be able to output or *source* a minimum of 400 μamps at 5 volts DC, and to receive or *sink* at least 1.6 ma at the logical 0 condition.

One important solid state alternative to TTL is CMOS (Complementary Metal Oxide Semiconductor). CMOS devices operate with logical 1 voltages between 3 and 15 volts DC (accuracy of the supply is not essential). CMOS devices are slower than TTL equivalents, but consume much less power.

Although it is beyond the scope of this book to discuss solid state logic circuit design, it is useful to point out several design concerns. First, the objective of solid state digital control circuit design is to minimize the *package count*, or number of solid state devices used in the circuit. This is done by trying to use the same type of logic gates (AND, OR, etc.) throughout the circuit. Also, the circuit should be simplified to the greatest possible extent using the Karnaugh map technique.

Logic gates can be wired together directly to form relay-ladder-type circuits, but care must be taken to assure that too many logic gates are not being driven by a single gate. *Fanout* is the maximum number of subsequent gate inputs which can be fed from any given logic gate output.

Solid state devices are fast, but require a finite time to switch states (i.e., to turn ON or OFF). *Propagation delay* is the time a gate requires to change its output state in response to a change in input state. Usually this delay is not important for the basic machine control function, but it may be important if the outputs are being read by a high-speed computer interface.

Solid state logic interfacing with a machine is not as straightforward as is the case with higher power electromechanical control systems. First, the TTL or CMOS gates require a low-level DC power supply to power the control circuit. The DC voltage need not be accurate, but it must be free from *filter* transient voltage spikes which can easily burn up solid state devices. This requires isolation of the logic gates from both inlet and outlet.

Interfacing solid state logic gates with the external world of relays, contactors, solenoids, and heaters is accomplished with two devices; *solid state relays* and *optoisolated input/output modules*, both of which are very similar.

Both are capable of switching 120 V AC power based on logic-level (TTL or CMOS) inputs. Fig. 9-20 shows an optoisolated triac circuit which will allow TTL power to switch 120 V AC, say to power a solenoid. (A *triac* is a solid state switch which passes AC current when turned on.) The entire circuit of Fig. 9-20, including the optoisolator, output triac, and *RC* protection network, is available as a potted module, compatible with PC boards, and is called a *solid state relay*.

Example 9-13 demonstrates the electronic control circuit equivalent of the relay ladder circuit originally developed in Example 9-7. The limit switches would best be replaced by proximity switches with solid state outputs, but this is not absolutely necessary. Note that the limit (or proximity) switches must switch logic-level voltage rather than 120 V AC or 24 V DC. The cylinder valve solenoids are isolated from the solid state components using an output module like that of Fig. 9-20. This control circuit is referred to as a *hard-wired control* because the machine sequence, once fixed by the control circuit, cannot be changed without physically altering the circuit elements.

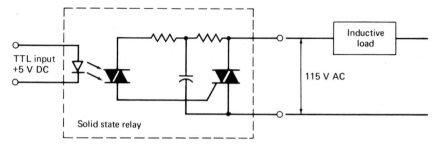

A solid state relay is basically a self-contained opto-coupled triac circuit which allows comp TTL logic circuits to switch AC power. The sensitive solid state circuit is protected from high-voltage line surges, etc. by the optoisolation feature of the relay.

Fig. 9-20. Solid state relay.

EXAMPLE 9-13
Implementing the Clamp-and-Work Sequence With Electronic Logic

Use the simplified Boolean control equations of Example 9-12 to design an electronic logic equivalent to the relay ladder control circuit.

The simplified Boolean control equations were found to be:

$$\text{SOL A} = \bar{M}$$
$$\text{SOL C} = 1\text{-LS} \cdot \bar{M}$$
$$M_1 = 2\text{-LS}$$
$$\text{SOL D} = M$$
$$\text{SOL B} = 3\text{-LS} \cdot \bar{M}$$
$$M_2 = 4\text{-LS}$$

These equations can be implemented with a hard-wired logic circuit made up of two two-input AND gates and a R–S (Set-Reset) flip-flop. The input switches are shown as standard limit switches; however, they might well be proximity switches (see Section 10.3) with logic-level outputs.

The outputs of the control circuit would be logic-level voltages (5 V DC), so an interface device, such as the solid state relay (Fig. 9-20) would be inserted between the logic circuit and the solenoid coil.

The logic circuit would be:

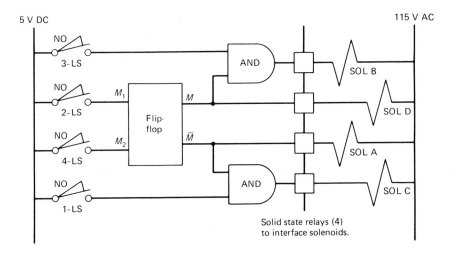

5 V DC

115 V AC

Solid state relays (4)
to interface solenoids.

9.5. PROGRAMMABLE CONTROLLERS

As discussed in previous sections, a machine control system can be purely electromechanical, using limit switches, relays, timers, etc. Up through the 1950s, the predominant mode of machinery sequence control was relay logic—implemented with relays. Control packages were large and bulky, usually packaged in heavy-duty, oil-tight electrical enclosures for protection from industrial environments.

Widespread use of semiconductor devices gave rise to hard-wired digital logic device control systems, initially with transistor-based logic gates. Early electronic gates were single function, individual components (i.e., AND, OR, NOR, etc.) which could be wired together to form the desired logic system. Medium and Large Scale Integration (MSI and LSI) brought about integrated circuit (IC) packages with many universal gates on a single chip, packaged in a single DIP.

Universal gates are NAND and NOR gates which can be manufactured on ICs with high gate densities. All other gate functions such as AND or OR can be obtained through combinations of NAND gates or NOR gates. It therefore often pays to transform the control logic into a network of all NAND or all NOR gates, using the identities discussed previously. However, it is beyond the scope of this book to consider this subject further.

In all solid state electronic control systems, the control logic is mechanized or implemented at very low (TTL or CMOS) power levels, therefore it must be interfaced with the real world of motors, relays and solenoids by means of input/output (I/O) modules. The solid state hard-wired systems are also

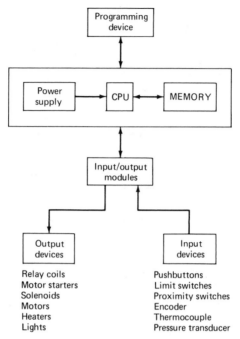

Fig. 9-21. Elements of a programmable controller.

fixed once built, thus dedicated to repeating a single control cycle over and over.

The advent of microprocessors in the 1970s led to the design of *programmable controllers* (PCs), essentially universal machine control or machine sequencing systems capable of transforming large numbers of control inputs into the desired control outputs with the *capability of easily changing the intermediate control logic*. Fig. 9-21 shows the typical PC structure.

Most PCs can be programmed in relay ladder logic format, and allow the user to place the equivalent of relays and counters into a ladder format exactly as they would with electromechanical components. The control outputs can be relay coil equivalents with contacts used elsewhere in the ladder, or output loads which represent motors or solenoids. Alternatively, most PCs can also be programmed by writing the desired Boolean algebra equations relating control inputs and control outputs.

The PC implements the desired control logic by running through all of the lines of the relay ladder equivalent and:

1. Looks at the state of the actual control inputs (i.e., is 1-LS open or closed, etc.);
2. Plugs the value or state of the input into its appropriate Boolean equation; and, if the equation is TRUE,
3. Actuates the output specified as the dependent variable of the Boolean equation

Input/output modules are available for PCs which interface all types of standard control input and output devices. Typical inputs include 120 V AC, 28 V DC, and logic-level DC voltages (for example, the output from a computer). Typical outputs include switching 440, 220, 120 V AC at various current levels; and 120, 28, and logic-level DC voltage.

Typical cycle time for the PC to read all control inputs, and perform the required logic to output is 20 msec.

9.6. MICROPROCESSOR AND MICROCOMPUTER CONTROL

The programmable controller described in the previous section is capable of even the most complex machine sequencing problems. In addition, many PCs can also perform arithmetic functions and route absolute measurement data to data collection devices (such as another computer). Why, then, consider a microprocessor or microcomputer control for a machine?

The first consideration is speed. The PC cannot respond to interrupts as can a microprocessor. The PC must complete one cycle of "looking" at inputs and commanding outputs before it recognizes the need for a change in machine status. The second reason for microprocessors is versatility. Generally the PC must be reprogrammed to change the machine control cycle. (However, some of the latest models of PC are programmable from a remote computer.) A microprocessor can access and execute several different control sequences (which would have to be stored in memory somewhere) upon the command to do so—even while the machine is running.

In general, a microprocessor is the CPU for a microcomputer. Whether it be on a single chip, a printed circuit board, or in a larger package, the only thing that discriminates a microcomputer from a microprocessor is the addition of memory, an oscillator to perform the timing function, and input/output circuits. A pure microprocessor control system is what a product designer considers when he is designing a "smart" product for mass production. In this case, unit costs may be saved by using a custom-tailored microprocessor system with the control program stored in mass produced ROM (Read Only Memory). However, the additional expense associated

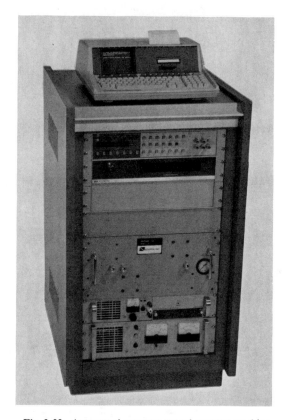

Fig. 9-22. Automated pressure transducer test machine.

with the more versatile microcomputer control system is insignificant for one-of-a-kind special machines.

The options available for implementing machine control with a microcomputer are many. The machine sequence may be time based or event based. Event-based machine sequencing can be implemented by interrupting the microcomputer activity when an event occurs (say a limit switch opens) and then jumping to the part of the program which tells the machine what to do next. Or, the computer can periodically check the limit switch to determine if it is open or closed, much the same way a PC scans the control inputs.

Because of its great speed, the computer may also perform other functions besides machine sequencing, "in-between" machine control commands. Typical functions include acquisition of data (inspection data), data manipulation, data transfer and storage, data or machine status readouts to the operator, etc. *Packaged microcomputers* are more expensive than single-board computers, but offer the designer a great deal of latitude in terms of mass data storage, data printout or recording, and ease of operator interface. It is often desirable to control a machine with a microcomputer capable of being programmed in high-level computer language, such as BASIC. This allows the machine to be fine tuned, or even have the control sequence changed, by the customer's engineers, rather than calling the builder in for modifications.

It is also often convenient to use a manufacturer's standard peripheral devices, such as *x-y* or *x-t* recorders to provide hard copy records for production engineering or for quality control. Fig. 9-22 shows a pressure transducer test machine which is controlled with an off-the-shelf Hewlett Packard HP-85 computer. The built-in strip chart allows the pressure transducer performance (voltage output versus pressure input) to be plotted in a format which not only shows the transducer output, but also prints the acceptable error limits for the particular transducer being tested right on the performance printout. This hard copy is used as official quality certification of that particular instrument.

The test machine can be programmed on the factory production floor to input pressure in any steps desired up to any range desired. Thus, 100 psig transducers can be tested in 10 psig incremental pressure inputs, or 1000 psig transducers may be tested with 100 psig incremental inputs. The operator simply "tells" the computer which instrument he is testing, and the test conditions desired. Communication with the machine is in BASIC. Use of off-the-shelf computers for machine control is becoming more popular because of this versatility. The only problem to date is that most "personal computers" sold today are packaged more for laboratory or office environments than they are for harsh industrial environments.

10. SENSOR PRINCIPLES

Sensors used in the design of automatic machinery fall generally into the following two categories, based on the manner in which they provide machine status information to the machine control system.

1. *Event sensors.* These sensors detect the occurrence of a specific event (e.g., position, temperature, light, pressure, time interval) and indicate the event with a digital (ON/OFF) output, usually a switch closure. These sensors are always *digital* in the sense that they are either ON or OFF.

2. *Continuous sensors.* These sensors measure physical variables such as position, velocity, acceleration, temperature, pressure, force, etc., and provide an output representative of the magnitude of that variable at any given time. These sensors may be either *analog* or *digital* devices.

As described in Chapter 9, the basic machine control requirement is usually machine operational sequence. In order to implement sequential controls, sensors are required to indicate to the controller that a certain machine operation is complete and to initiate the next operation. Clearly, sensors of the event type which provide either ON or OFF indications are sufficient to control the machine sequence.

Machinery control systems also provide a machine protection function in addition to the control function. In general, event type sensors can be used to detect an unsafe machine condition, and open or close a switch in response to that condition. Typical sensors are bimetallic temperature switches, pressure switches, or photoelectric switches.

Continuous sensors are generally used in the following applications:

1. Closed-loop servosystems, where the actual magnitude of machine output (position or velocity) is required for comparison with desired magnitude of machine output so that corrective action may be taken. (Note that the use of a closed loop drive in a machine may improve machine accuracy or reliability, but does not in any way contribute to machine operation sequencing.)

2. Measurement systems, where the absolute value of some physical property of the machine or the work is required as data to be processed, indicated, recorded, and/or otherwise stored.

3. Process control subsystems of a machine, where maintenance of a certain environmental condition is necessary, such as temperature control in an on-line oven, or pressure control in an on-line autoclave.

4. Machine operation sequencing in conjunction with a threshhold circuit or device which ultimately gives an ON/OFF signal when the absolute value of some machine variable reaches a preset value. The combination of continuous sensor and threshold circuity, however, is indistinguishable in operation from an event type sensor.

10.1. ELECTROMECHANICAL SWITCHES

Machine control sequencing generally requires event type sensors which indicate an event by means of a switch opening or closing. Because of the importance and widespread application of these switches in their varying configurations in machine design, a general discussion of switches and switch contacts follows.

10.1.1. Mechanical Contact Switching

In order for a switch to provide meaningful control data, it must pass an electrical current with a virtual short circuit (therefore no voltage drop) in the ON state, and break that current flow (maintaining applied voltage across an open circuit) in the OFF state. The location at which current is interrupted in a mechanical switch is known as the *contacts*.

The basic problem with mechanical contact switches is that current, once flowing through the closed contact pair, has a tendency to keep flowing when the contacts are opened. The result is an arc, or spark, jumping from one contact to the other as the contacts separate. The greater the amount of current flowing through the contacts, and the higher the voltage across the open contact pair, the greater is the tendency to arc.

As the contacts are opened, arcing across the air gap results in spark erosion of the contact surfaces as well as contact oxidation. When the contacts are closed again, the arc-eroded contact surfaces offer slightly higher resistances than they did on the previous closing.

Higher surface contact resistances result in $I^2 R$ heat dissipation at the contact interface. In fact, instantaneous temperatures at contact points of asperities (microscopic peaks of the spark eroded surface) are high enough to melt the contact material, welding the contacts together. When the switch is opened, microscopic chunks of contact material are literally pulled from one or the other contact, further aggravating the surface erosion problem.

It is not difficult to understand, then, that all mechanical switch contacts must fail sooner or later by virtue of the degradation inherent in the switching operation itself. To maximize contact life, switch designers attempt to minimize contact resistance (thus heating) by using oxidation-resistant metals (gold, platinum) and by providing contact pressure in the ON state to squeeze the contact interface, increasing contact area and breaking through any oxide film.

The invention of the snap-acting switch in 1932 by P.K. McGall was a major breakthrough in the field of electrical control. Then-new beryllium copper was used as a spring which was able to make and break switch contacts quickly, repeatably, and with good force (hence "snap-acting") with only a small amount of actuator or plunger travel. Snap-acting contacts are used today almost exclusively in electromechanical relays, limit switches, and timers.

Switch contacts are closed (or opened) in response to a linear or rotary motion of the switch *actuator*. The actuator travels through a distance, known as *pretravel*, to the switch *operating point*. During pretravel, the actuator is storing energy in the switch spring. At the operating point, the contacts start to move, and indeed, if no further actuator travel took place, would snap into their closed position. However, the actuator usually overrides the spring by an amount known as *overtravel*, and holds the contacts closed with some pressure. As the actuator force is removed, the actuator reverses direction to the *release point*, at which the contacts snap back open. The release point (on opening) is usually closer to actuator initial position than the operating point (on closing), and the difference between the two is called *differential travel*. The distance through which the movable contact travels is never greater than $\frac{1}{8}$ inch. Actuator travel can vary greatly depending on the type of switch.

Switching Resistive and Inductive Loads.   There are two basic types of electrical loads which must be switched by machine control systems: resistive and inductive. Resistive loads include lights and heaters, while inductive loads include motors and solenoids.

Contacts carrying DC current to purely resistive loads have a tendency to arc when the circuit is broken which is proportional to open circuit voltage and amount of current being carried through the circuit. If a pair of contacts carrying AC power is opened at precisely the time the AC voltage passes through zero, no arcing will occur. In fact, the principal of *zero voltage switching* takes advantage of this fact, and is the preferred technique of AC heater temperature control. In the worst case, AC power to a resistive load will be switched at its peak voltage and will arc according to that voltage.

The problem with switching inductive loads carrying DC power is that an abrupt opening of the circuit results in a very high voltage surge which not only aggravates arcing, but also is potentially damaging to any other electronic components in the same circuit.

Another problem associated with switching motors, relays, and solenoids is that they exhibit high inrush currents upon startup. These initial high currents can result in contact welding as discussed above, and so need sufficient spring force to break the contacts.

Contact arcing can be suppressed magnetically by placing a coil in close proximity to the opening contacts with a magnetic field which opposes that of the arc itself. Contact arcing in DC inductive load circuits can also be suppressed by inserting resistors, capacitors, and/or diodes in parallel with the contacts and the inductive load. These components allow the inductive power surge to be dissipated at circuit locations other than across the opening contacts.

Contact Arrangements. The state of a switch's contacts in the unactuated mode determines whether a switch is normally open (NO) or normally closed (NC) with respect to a single circuit. A NO contact pair will close (pass current) or *make* upon switch actuation. A NC contact pair will open (stop current flow) or *break* upon switch actuation.

The number of *poles* associated with a switch are the number of independent circuits which the switch makes or breaks at one actuator position.

The *throw* of the switch is the number of circuits which can be switched by a single pole. For example, a single-throw switch might make a circuit in one switch position, and break that same circuit in another position. A double-throw switch might have circuit A closed and circuit B open in one switch position, and vice versa in the other switch position.

Switch *break* is the number of contact pairs which are opened or closed in each independent circuit as the switch moves from one position to the other. Fig. 10-1 shows schematic diagrams of contact arrangements depicting NO, NC, poles, throws, and breaks.

Figure 10-2 shows the ANSI contact symbols for a number of switch contact arrangements. The *form* is identified by a letter as shown in the figure. The number of poles, or the number of separate circuits which can be switched, is identified by a number preceding the form letter. Thus a 4A switch would be a four-pole NO switch. If it were a relay, then the relay coil would control four pairs of NO contacts. A 2C relay would have two pairs of

Single-pole, single-throw, single-break, normally open contacts.	
Double-pole, single-throw, single-break, normally open contacts.	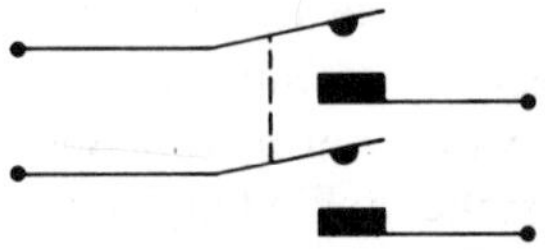
Single-pole, single-throw, double-break, normally open contacts.	
Double-pole, single-throw, double-break, normally open contacts.	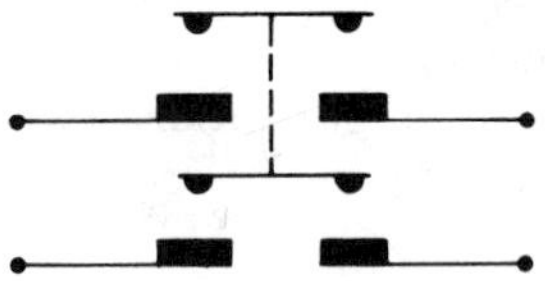
Single-pole, double-throw, single-break, one pair NO, one pair NC contacts.	
Double-pole, double-throw, single-break, two-pair NO, two-pair NC contacts.	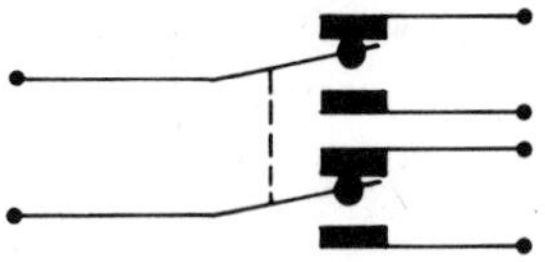
Single-pole, double-throw, double-break, two-pair NO, two-pair NC contacts.	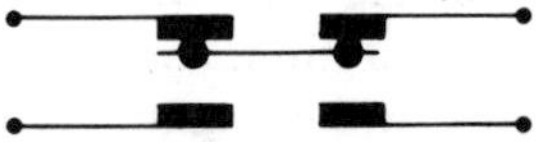
Double-pole, double-throw, double-break, four-pair NO, four-pair NC contacts.	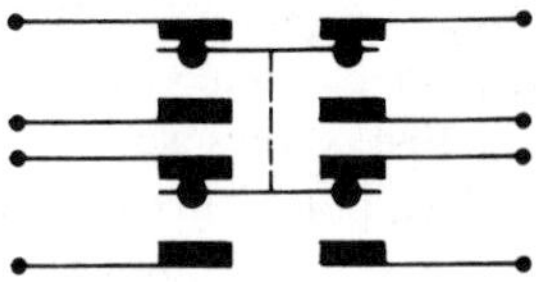

Fig. 10-1. Electromagnetic contact arrangements.

(a) ANSI symbols for electromechanical contacts

Form	Description	Symbol	Form	Description	Symbol
A	Make or SPSTNO		K	Single pole, double throw center off or SPDTNO	
B	Break or SPSTNC		L	Break, make, make, or SPDT (B-M-M)	
C	Break, make or SPDT (B-M) or transfer		M	Single pole, double throw, closed neutral, SP DT NC	
D	Make, break or make-before-break, or SPDT (M-B), or "Continually transfer"		U	Double make, contact on arm, SP ST NO DM	
E	Break, make, break, or break-make-before-break, Or SPDT (B-M-B)		V	Double break, contact on arm, SP ST NC DB	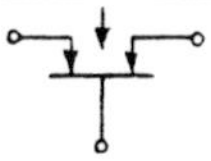
F	Make, make SPST (M-M)		W	Double break, double make, contact on arm, ST DT NC-NO (DB-DM)	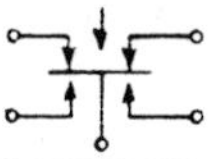
G	Break, break or SPST (B-B)		X	Double make or SP ST NO DM	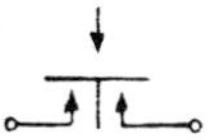
H	Break, break, make, or SPDT (B-B-M)		Y	Double break or SP ST NC DB	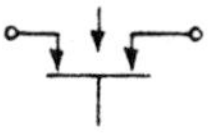
I	Make, break, make, or SPDT (M-B-M)		Z	Double break, double make SP DT NC-NO (DB-DM)	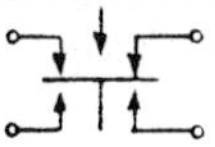
J	Make, make, break, or SPDT (M-M-B)				

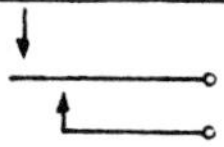

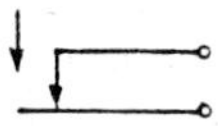

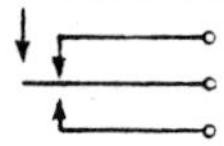

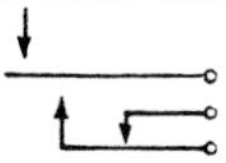

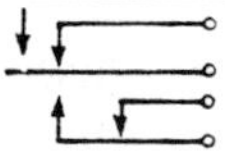

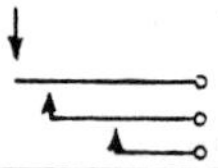

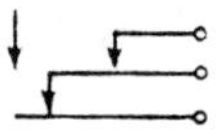

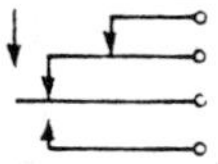

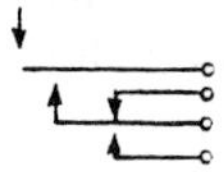

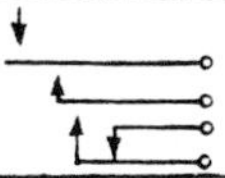

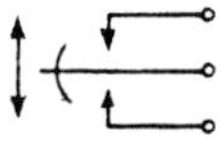

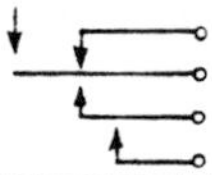

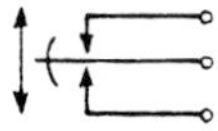

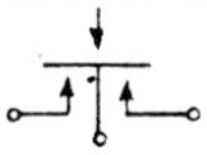

(a)

(b) JIC symbols used in relay ladder diagrams

Form	Symbol	Description
A	—| |—	Normally open
B	—|/|—	Normally closed
C	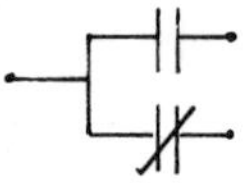	Transfer (Break before make)
D	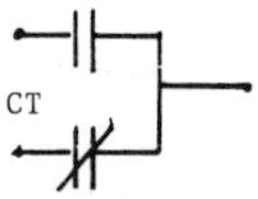 CT	Continuous Transfer (Make before break)

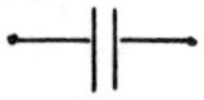

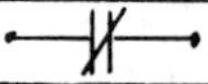

(b)

Fig. 10-2. (a) ANSI symbols for electromechanical contacts, and (b) JIC symbols used in relay ladder diagrams.

transfer contacts controlled by the relay coil. Transfer contacts are double-throw contacts because they are normally closed for circuit A (NO for circuit B) in the unactuated state, and vice versa in the actuated state. A form 2C relay could also be called a double-pole, double-throw switch.

Also shown in Fig. 10-2 are the JIC (Joint Industry Council) standard symbols for some contacts. This book uses the JIC symbols because they are most commonly used in relay ladder diagrams (Section 9.2.3).

10.1.2. Manual Switches

Manually actuated switches are of the *momentary contact* or *maintained contact* types. Momentary contact switches include pushbutton and toggle switches, and this type of action is widely used to start machines. The relay holding circuit described in Section 9.2.3 allows momentary contact switches to both start and stop machine operation.

Maintained contact switches include pushbutton, toggle, sliding, and rotary types. The contacts are transferred by the actuator, and remain transferred after the operator removes his hand. This is true whether or not the switch actuator returns to its original position. A positive action must then be taken to transfer the contacts to their original state. This action may be a reverse motion from that originally applied, or a repeat of the original motion (common in maintained contact pushbuttons). Maintained contact manual switches often have integral indicator lights to show whether they are in the OFF or ON state.

10.1.3. Limit Switches

Limit switches are perhaps the most common linear position sensors used in automatic machine design. A limit switch is designed to be mechanically actuated when a machine member reaches a particular position. The machine member physically contacts the limit switch actuator and switches the contacts, normally one to four poles of Form A, B, or C. As discussed in Chapter 9, the limit switch can be used to control machine operations sequencing. Limit switches also provide machine protection functions, shutting down the machine when machine members or work are at a location where they should not be.

Industrial limit switches are usually enclosed in order to protect the switch from dust, water, and human abuse. Standard duty enclosed limit switches are rugged; however, *heavy duty* limit switches are available which are oil-tight, corrosion resistant, and generally more rugged devices. Heavy-duty limit switches should undergo 20 million cycles without failure under moderately severe industrial usage.

TABLE 10-1.   NEMA Switch Enclosures.

Non-Hazardous Locations

Type 1
: For indoor use primarily to provide a degree of protection against contact with the enclosed equipment.

Type 3
: For outdoor use primarily to provide a degree of protection against windblown dust, rain, sleet, and external ice formation.

Type 3R
: For outdoor use primarily to provide a degree of protection against falling rain, sleet, and external ice formation.

Type 4
: For indoor or outdoor use primarily to provide a degree of protection against windblown dust and rain, splashing water, and hose-directed water.

Type 4X
: For indoor or outdoor use primarily to provide a degree of protection against corrosion, windblown dust and rain, splashing water, and hose-directed water.

Type 6
: For indoor or outdoor use primarily to provide a degree of protection against the entry of water during occasional temporary submersion at a limited depth.

Type 12
: For indoor use primarily to provide a degree of protection against dust, falling dirt, and dripping noncorrosive liquids.

Type 13
: For indoor use primarily to provide a degree of protection against dust, spraying of water, oil, and noncorrosive coolant.

Hazardous Locations

Type 7
: For use indoors in locations classified as Class 1, Groups B, C, or D by the National Electrical Code.

  Group B—Atmospheres containing hydrogen or manufactured gas.
  Group C—Atmospheres containing diethyl ether, ethylene, or cyclopropane.
  Group D—Atmospheres containing gasoline, hexane, butane, naptha, propane, acetone, toluene, or isoprene.

Type 9
: For use in indoor locations classified as Class II, Groups E, F, or G, as defined in the National Electrical Code.

  Group E—Atmospheres containing metal dust.
  Group F—Atmospheres containing carbon black, coal dust, or coke dust.
  Group G—Atmospheres containing flour, starch, or grain dust.

Table 10-1 lists the common NEMA type enclosures for limit switches, and the conditions against which they must protect switches.

Figure 10-3 shows the common limit switch actuators used in the machinery design. Actuator types are selected primarily on the basis of where the limit switch might be located with respect to the controlled machine member. It is important to install the limit switch in a location and position such that:

- The moving machine member will not destroy the limit switch if, for some reason, the limit switch fails to work (i.e., the cam overrides the switch actuator).
- The limit switch is accessible for maintenance.

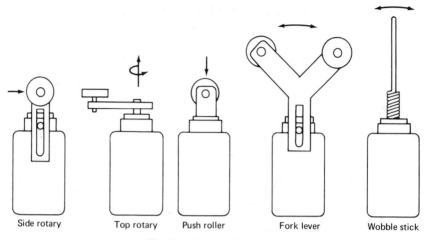

Side rotary   Top rotary   Push roller   Fork lever   Wobble stick

Fig. 10-3. Limit switch actuators.

- The limit switch is protected from accidental actuation (such as an overhanging workpiece, etc.).
- Chips, moisture, grease, or oil accumulation does not take place on the actuator (which can affect machine accuracy at switching point).

The limit switch is usually attached to a stationary machine element. A *trip dog* or *cam* is mounted on the moving machine element, and as the moving element passes by the limit switch actuator, the cam actuates the switch. The cam position is always made adjustable (not the switch itself) so that switching may take place at different moving element positions.

Figure 10-4 illustrates proper cam design for mechanically actuated limit switches. Improper cam design can result in greatly reduced switch life.

The primary reason for premature limit switch failure is repeated high impacts due to improperly designed cams, particularly at high speeds. First, the leading edge of the cam must contact the actuator roller at the proper angle. Fig. 10-4(a) shows the cam leading edge parallel to the switch arm within ± 15°. This cam *overrides* the roller, and in doing so, it must not rotate the switch arm beyond the recommended overtravel for that switch. If the overtravel limit is exceeded, the switch or arm itself may be broken if the arm is rigid. If the arm is flexible, it may be overstressed, and eventually fail in fatigue.

For cam speeds of less than 50 feet per minute, a 45° straight cam angle as shown in Fig. 10-4(a) is satisfactory. For cam speeds in the range of 50–200 feet per minute, the cam angle should be reduced so that the fast motion does not overaccelerate the actuator, resulting in excessive contact bounce. Fig.

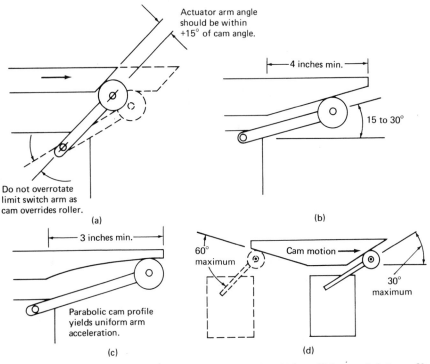

Fig. 10-4. Limit switch cam design: **(a)** for speeds less than 50 fpm; **(b)** for speeds between 50 and 200 fpm; **(c)** for speeds greater than 200 fpm. **(d)** Constant-acceleration cam design.

10-4(b) shows a minimum straight cam length of 4 inches. For speeds up to 400 feet per minute, it is necessary to control actuator acceleration by using a curved cam. A parabolic cam shape [Fig. 10-4(c)] results in a constant actuator acceleration. Note that the actuator arm is adjusted in both Fig. 10-4(b) and Fig. 10-4(c) so that the leading edge of the cam is parallel (within ±15°) to the actuator arm.

Just as it is important to avoid excessive impact of cam on actuator, so is it important to avoid rapid spring back as the cam overrides the actuating roller. Under no circumstances should the lever be allowed to snap back. Fig. 10-4(d) shows the trailing edge of the cam not exceeding 60° between the lever arm and the cam surface. Note that the use of an override cam will result in the actuation of the switch on the return stroke unless a one-way lever is used.

One-way cam actuation can be accomplished with a *knee-joint* actuator arm as shown in Fig. 10-5(a) or a *hinged dog* trip lever as shown in Fig. 10-5(b). Both one way designs should be limited to moving machine element speeds of less than 50 feet per minute.

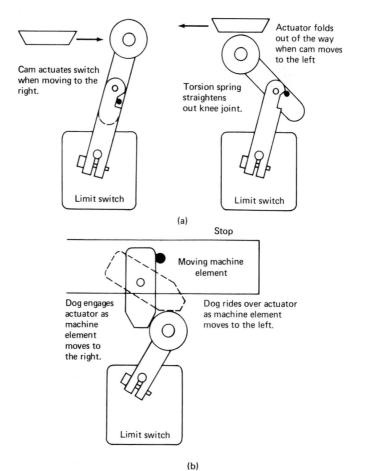

Fig. 10-5. (a) Knee-joint actuator for one-way limit switching. **(b)** Hinged dog actuator for one-way limit switching.

Impulse Limit Switches. Impulse limit switches are available which produce a switching signal only while the plunger travels in. The contacts switch back when the end of plunger travel is reached, and remain OFF during return travel of the plunger. This switching action overcomes some control system design problems where a cam sitting on a limit switch will not allow subsequent operations to take place.

The duration of the switching impulse is inversely proportional to actuation speed. Generally, 50 feet per minute is the maximum speed for reliable operation. Positioning of the impulse switch relative to its actuating cam is critical because the plunger travel must be enough to complete the switching pulse, but not so much as to cause overtravel damage.

As shown in Example 9-8, the impulse switch can simplify a control system, but as shown in Example 9-7, the same sequence can be obtained with standard limit switches and holding relays.

10.1.4.   Reed Switches

The *reed switch* or *reed relay* consists of two or more thin metal strips (reeds) enclosed in an evacuated hermetically sealed glass capsule. The reeds overlap and can be closed by moving a magnet close to the reeds, and opened by moving the magnet away. Alternatively, the reeds can be closed by energizing a nearby coil (reed relay). Fig. 10-6 shows a NO reed switch. A coil wound around the switch creates a reed relay, and when energized, the reeds snap together and remain closed until the coil is de-energized. To create a NC relay contact, the reeds are biased into the closed position with a permanent magnet. A reverse polarity coil cancels the permanent magnetic field when energized, and the reeds snap open.

Reed relays can make up to 30 amps and break 3 amps of 120 V AC power. They can make or break 3 amps of 24 V DC power or 1 amp of 125 V DC power or 1 amp of 125 V DC power. Reed switches are not generally used to switch large inductive loads directly (i.e., motors), but often are used to switch power to control relay coils, which in turn switch power to motors.

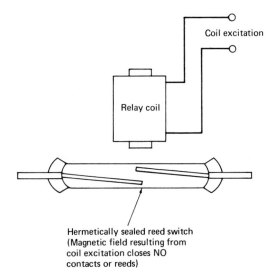

Fig. 10-6. Normally open sealed reed switch.

10.1.5.   Pressure Switches

Pressure switches are used in automatic machinery as sequence control devices and as machine protection devices. A pressure switch is used in a machine control system in exactly the same way as a position sensing limit switch is used. Snap-acting contacts open (or close) when a preset pressure is sensed by the pressure switch.

The pressure switch is not unlike the limit switch in construction, indeed, some pressure switches employ a basic plunger-actuated limit switch in their design. What must be added to a limit switch to make it a pressure switch is a pressure sensitive subsystem which outputs an actuating stroke (proper force and distance) to open and/or close snap-acting contacts. The key to designing a useful pressure-sensing subsystem is to make the pressure at which the contacts switch adjustable.

Pressure switches are available in most NEMA-type enclosures for protection from various environments. The switch assembly has one or two NPT ports to which the pressure being sensed is connected, and can be used with liquid or gas media. If switching is to occur at a given *differential pressure*, then fluid lines at the two pressures (high and low) are connected to the switch. If switching is to occur at a set gage pressure (set point) then only one fluid line is connected to the switch.

Pressure switches use either pistons, bellows, or diaphragms as the pressure-to-stroke switch actuator. Bellows and diaphragms are most common because their static seals can prevent fluid leakage very reliably. The sensing element (diaphragm or bellows) is usually biased with an adjustable spring load, so that the pressure at which the switch actuates can be adjusted.

Depending on the direction of initial spring bias, the pressure switch can be made to actuate when the line pressure reaches the set point pressure from below (i.e., pressure increasing) or from above (i.e., pressure decreasing). The contacts remain closed (open) until the line pressure reverts to a value known as the *reset point* or *reactuation point*. The reset point is usually different from the set point. In the case of a switch which actuates on pressure rise, the reset point will be lower than the set point. This difference in pressure between the set and reset points is known as the *differential* or *dead band*. In some pressure switch models, the dead band is adjustable (i.e., the set point and reset points are adjustable independently).

Use of pressure switches in fluid power automation systems can often simplify component placement problems in designing the machine. Fig. 10-7 shows a cylinder with pressure switches sensing the cylinder inlet and outlet pressures. This arrangement can replace two position sensing limit switches and the attendant problems of switch placement and cam design. If pressure

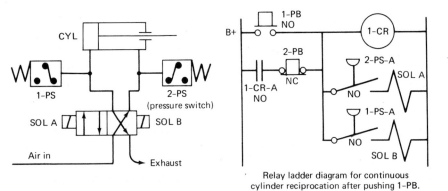

Relay ladder diagram for continuous
cylinder reciprocation after pushing 1–PB.

Fig. 10-7. Use of pressure switches to replace limit switches in pneumatic systems.

switch 1 is set to actuate at line pressure, then a switch closure will result when the piston rod is fully extended and cap end pressure rises to line pressure. Similarly, switch 2 will actuate when the piston is fully retracted and pressure in the rod end reaches line pressure.

Note, however, that the piston stroke can not be adjusted by changing a cam position as it could be if limit switches were used. Unless external stops are set to limit piston rod extension, the total stroke of the cylinder will be the only choice available. Also, the possibility exists of accidental switch actuation as a result of a line pressure surge.

The primary advantage to the system of Fig. 10-7 is that the pressure switches themselves may be located somewhere other than on a machine element. However, very long pressure sensing lines may result in time delays and should be avoided.

The system of Fig. 10-7 can also be used as a *load sensing* control scheme, a feature not possible with limit switches. For applications where it is necessary to limit applied force, the pressure switch sensing the cylinder cap end pressure will have its setpoint adjusted to switch at a pressure equal to the maximum force divided by the piston area. When pressure in the cylinder rises to the point where the applied force equals the maximum allowed force, the switch will close, and by virtue of a properly designed control system, retract for the next machine cycle.

10.1.6. Electromechanical Timers

Manual switches, electromechanical control relays, limit switches, and pressure switches are mechanically actuated devices which require no external power other than that in the control ladder to generate control or sequencing signals. Furthermore, they switch control or line power directly through their contacts and are relatively easy to troubleshoot (i.e., there are not many failure modes). For this reason, these electromechanical components retain a certain popularity relative to their solid state equivalents among control engineers.

Electromechanical timers, on the other hand, do not hold any unique advantages (real or imagined) over their solid state equivalents, and indeed, solid state timers are more reliable, repeatable, and faster acting than previous designs. The reason that solid state timers were quickly accepted is that they are one-for-one replacements for their electromechanical predecessors. Both require a source of external power, both start or stop their timing intervals in response to line or control power inputs, and both switch the same levels of load power at their output terminals.

The basic timer function is to open or close switch contacts on a time basis rather than an event basis. The timing of switch openings or closures is adjustable from several seconds to many hours. It is the adjustable time interval switching feature which distinguishes a *timer* from a *time delay relay*. The time delay relay simply opens or closes contacts at some time interval after the relay coil is energized.

Timers are known as either *reset timers* or *repeat cycle timers*. Reset timers open or close their contacts upon timer energization. The contacts remain transferred for a predetermined time interval, after which they revert to their initial state. The input signal must be removed, then reapplied to the reset timer in order for the timed contact closure to begin again. If the input power is removed before the timer "times out," the timer resets and the output contacts switch back to their de-energized state.

The most common electromechanical reset timers use a synchronous gear motor as the time reference. Upon energization at the timer input, the motor starts, and a solenoid actuates a clutch which connects the motor to a shaft with an adjustable cam. When that cam (position is set as the time adjustment) rotates through the preset angle (or time) it opens (times out) the output contacts. The motor remains connected to the timing shaft as long as input power is applied to the clutch solenoid. When input power is removed, the clutch opens, and a spring resets the timing shaft, making the timer ready for another cycle.

Electromechanical repeat cycle timers use a synchronous gear motor as the timing source to drive one or more circular cams which alternately open and close switch contacts as the timing shaft rotates. Repeat cycle time is limited to the output speed of the motor, and the cycle stops at the point of motor de-energization.

10.2. Solid State Switches

It has been shown in Chapter 9 that electromechanical devices such as switches, relays, limit switches, timing motors, etc. can be put together to create any desired machine control or sequencing system. The entire control system can work with commonly available power—120 V AC single phase. The devices are rugged, not particularly temperature sensitive, and circuits are easy to design and troubleshoot because they are easy to understand.

Why then should control system designers want to use solid state control components at all? There are three major reasons: speed, reliability, and cost. Solid state systems offer much faster control (switching) speeds, greater component reliability, and lower component cost, although their application is more complex and requires considerable care to protect the system from industrial environments. In addition, solid state controls, particularly solid state switches, are directly compatible with computer or microprocessor control systems.

Solid state switches consist of multiple layers of semiconductor material configured such that the output terminals are nonconducting (open) until relatively low power is applied to the switch input terminals. The output terminals then become conducting (close) without the inherent unreliability of moving contacts. Each solid state switch is equivalent to a single-pole, single-throw (SPST) mechanical switch, except that actuation is accomplished electrically rather than mechanically. Solid state switching takes place in the microsecond-to-nanosecond range, as opposed to milliseconds required for mechanical contact switching.

Semiconductor switches do have a finite "contact" resistance, even in the conducting or ON mode. A 40 amp switch might typically dissipate 50 watts, so it is important that these devices be properly heat sinked and cooled. Unlike metallic contacts, semiconductor switches are very temperature sensitive, both in terms of poor performance at elevated temperatures and failure tendency due to thermal cycling (i.e., fatigue due to expansion and contraction).

Solid state switches are also highly intolerant of excessive voltage and current. It is therefore essential that switching circuits be protected from high inrush currents and voltage surges encountered when switching OFF inductive loads. It is not uncommon to "step up" from solid state control circuitry using the solid state switch or relay to activate an electromechanical relay, which in turn switches the load.

Solid state switches are now widely used as integral output switches for sensors which are not mechanically actuated as are standard limit switches. Major examples are proximity switches, photoelectric switches, and timers.

These sensors can be applied in control systems exactly as their electrome-chanical counterparts might be, as long as the switched loads do not exceed the output rating of the device.

10.3. PROXIMITY SWITCHES

Proximity switches are linear position sensors which can be applied in automatic machinery in the same manner as limit switches, but with one important distinction: physical contact between the switch and a machine element is not required to open (or close) the switch. The *target* or machine element approaches the proximity switch sensing area in either a *head-on* mode [Fig. 10-8(a)] or a *slide-by* mode [Fig. 10-8(b)]. When approached by a target in the head-on mode, the switch will open (close) contacts at a very repeatable target distance from the sensor. Switching accuracy of 0.001 inch or better is not uncommon. Unfortunately, the head-on mode of sensing must be avoided where possible to minimize the chance of sensor damage if the switch fails to stop target motion. If used in the head-on mode, the proximity switch must be protected by a hard stop, such as that shown in Fig. 10-8(c).

In the slide-by mode, the proximity switch will actuate when a given area of target is covering the sensor face. Just what this area is depends upon air gap between target and sensor, and switch sensitivity.

The proximity switch is not a simple device. It generally requires an external source of excitation power, an analog sensor which responds to the proximity of a material, a circuit to transform the analog proximity signal to a repeatable switching signal, and an output switch, usually solid state. In spite of the complexity, proximity switches are available in completely self-contained packages ranging in size from $\frac{3}{8}$ inch diameter to three inches in diameter. The same NEMA enclosure options available for limit switches apply to proximity switches.

The most widely used proximity switches are of the inductive type. The sensor or probe projects a high-frequency (200,000 Hz) electromagnetic field in the space in front of the sensor. When a *conductive* target enters the field, eddy currents are induced in the target by the field, and the probe detects these eddy currents. The Eddy Current Killed Oscillator (ECKO) principle is utilized, whereby eddy currents reduce an internal oscillator output to zero to create the switching signal. This internal probe signal can be used (with appropriate support electronics) to actuate different types of switch out-puts—relay contacts, time delayed relay contacts, solid state switching, or logic-level pulse outputs.

Inductive-type proximity switches are limited to use with conductive (usually metallic) targets. Sensing ranges vary from 0 to 1 inch. Switching capability and speed depends upon the type of switch output. Because of the

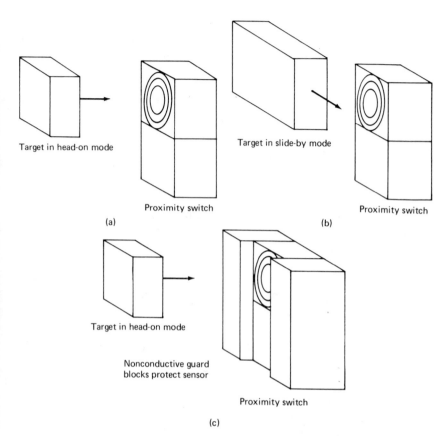

Fig. 10-8. (a) Proximity switch mounted in head-on mode. **(b)** Proximity switch mounted in slide-by mode. **(c)** Protecting the switch in the head-on mode.

high-frequency fields involved, a sensor must be shielded first from metallic machine elements, and secondly, from nearby sensors.

Other proximity switches are available commercially which utilize different operating principles:

Capacitive-type proximity switches are available which obtain their switching signal based on the air gap between the sensor and *nonconductive* materials.

Magnetic reed switch sensing schemes are used for proximity sensing by placing a permanent magnet on the travelling target. When the magnet comes into proximity to the stationary switch, the reeds close.

Hall Effect switches use Hall Effect devices which output a voltage proportional to magnetic field. A magnet mounted on the target creates a Hall voltage as it nears the sensor. This signal is amplified and when it reaches a threshold value, solid state switch outputs are actuated.

Ultrasonic proximity sensors operate on the principle of reflected ultrasound. The time between signal output and return of the reflected signal is proportional to distance between the sensor and the target. Developed initially as automatic focusing devices for cameras, these sensors operate over the range of 1–20 feet. Accuracy of switching is not as repeatable as that of the previously discussed sensors.

Proximity probes of all types are usually sealed, cylindrical or square devices with leads protruding from the non-active end. Heavy duty-proximity switches are often square assemblies, much like mechanical limit switches but without the *actuator*.

Ring-type proximity sensors are available which actuate (switch) when objects pass through the ring. Coin counters or small parts counters are examples of ring proximity switch applications.

10.4. PHOTOELECTRIC SWITCHES AND CONTROLS

Photoelectric controls are noncontact position-sensing devices which output switch closures in response to:

1. Interruption of a light beam by a workpiece or machine element;
2. Reflection of a light beam back to its source by a workpiece or machine element passing in front of the projected beam.

Figure 10-9(a) shows the typical set up of *light source* and *photoreceiver* for a *direct scan* or *through scan* photoelectric control system. When the light path between source and receiver is interrupted, an output in the form of a switch closure is obtained. Output contacts of Forms A, B, and C are widely available with ratings of up to 10 amps. AC devices can switch circuit loads such as relays, solenoids, or small motors directly. DC devices are usually used for logic-level switching; the output switches are solid state and switch DC currents on the order of 150 ma.

Direct scan devices can be used to count parts or work moving by on conveyors; as "light screen" safety systems to disable or shut down a machine when the screen is interrupted by an operator's hand; for on-line inspection of bottle fill levels; for break detection in long continuous webs such as paper,

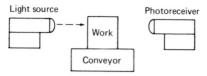

As work moves between light source and receiver, the light beam
is interrupted resulting in a switch closure (or opening).

(a)

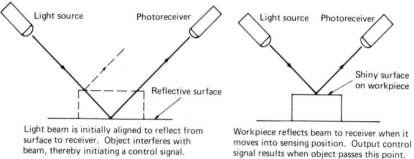

Light beam is initially aligned to reflect from
surface to receiver. Object interferes with
beam, thereby initiating a control signal.

Workpiece reflects beam to receiver when it
moves into sensing position. Output control
signal results when object passes this point.

(b)

Light beam from source to retroreflector and back to scanner is
interrupted by object, initiating a control signal.

(c)

Fig. 10-9. (a) Direct scan photoelectric sensing system. **(b)** Reflective scan photoelectric sensing
system. **(c)** Retroreflective photoelectric sensing system.

fabric, or plastic; as well as in many other applications. Through scanning
arrangements can be used with source/receiver separation distances of over
100 feet, although machinery applications rarely call for such a range.

Figure 10-9(b) shows a *reflective scan* arrangement, wherein the receiver
detects light reflected from the source by a work piece or part. Reflective scan
applications include liquid level sensing by reflection from liquid surface; part
or work detection (the part must have a reflective surface); part orientation
determination; web or strip positioning for cutoff; and many others. Web or
strip positioning usually controls continuous rolls in response to a dark

registration mark passing under the beam, causing the receiver to sense an interruption in reflected light. The control signal then stops the rolls, usually for purposes such as cutoff to length, pattern dyeing, printing, etc.

Figure 10-9(c) shows a *retroreflective* system, where both light source and photoreceiver are packaged in the same housing, known as a *scanner*. The retroreflective scanner can switch when the light beam from a fixed reflector is interrupted, or it can switch upon retroreflection from a passing workpiece or machine element. *Diffuse scanners* are also available which actuate switching upon diffuse reflection from a nearby rough surface. Such scanners serve as proximity sensors.

The light source for most industrial photoelectric control systems is the light-emitting diode (LED) which generates light in the infrared (IR) spectrum. Most industrial light sensors use silicon photodetectors which are sensitive to IR light and are economical. The IR beam is *modulated* or *chopped* at high frequency before leaving the light source package. If necessary, an optical lens focuses the beam as it leaves the source. The detector is tuned to the frequency of the source light, so it does not respond (switch) to background IR radiation. The photoreceiver detects light from the source, amplifies it, and switches its output contacts when a change in nominal conditions occurs. A receiver can be energized on light or energized on dark depending on its intended application.

Miniature photoelectric sources and receivers are available in which the light is transmitted to the source point, and received through fiber optic cables. This allows use of photoelectrics for parts counting or parts presence sensing devices in very tight machinery spaces.

Modulated IR light penetrates smoke, steam, fog, etc. fairly well, but source and receiver optics are often fouled by grease, dirt, and oil so that the transmitted signal is too weak to be effective. Photoelectric devices should always be mounted so as to avoid or be protected from dirty environments. The devices should also be selected with *excess gain G* appropriate to the environment. The excess gain for clean air is 1.0, that is, no special allowance need be made (although manufacturers rarely recommend an excess gain of less than 2.0 even for the cleanest air). The excess gain requirement increases to 2.0 for low contamination, to 10.0 for moderate contamination, and to 25.0 for high contamination.

Excess gain charts are usually presented for each source and receiver listed in manufacturer's catalogs; these charts plot excess gain versus working distance (between source and receiver). The closer the working distance, the higher the excess gain characteristic of the photoelectric device. Thus, if a source was located in an area of low contamination ($G = 2.0$), and the receiver in an area of moderate contamination ($G = 10.0$), the total excess gain for the source/receiver pair should be $2.0 \times 10.0 = 20.0$. The designer would then make sure that the working distance was less than that required for an excess gain of 20.0 for the particular devices used.

10.5. ROTARY POSITION SENSORS

The linear position sensors discussed in previous sections were basically presence sensors. It would be difficult to use these devices to accurately measure distance over distances on the order of feet as opposed to fractions of an inch. It turns out, however, that most very-long-stroke machine element travel is actuated by rotating drivers such as motors driving lead screws, chains, belts, or rack and pinion. As a result, the ultimate linear machine motion can be directly related to motion of a rotating member (shaft, gear, etc.) somewhere in the machine. Devices which measure rotational position can be used to indicate linear position, or velocity, by relating position changes to time. Table 10-2 summarizes the types of rotary position sensors used in automatic machinery.

10.5.1. Rotary Encoders

The most popular angular position sensor for automatic machine use is the digital rotary encoder. Output signals are DC voltage pulses of whatever level is required by the interface electronics which utilize encoder pulse outputs for a control function. The three general types of angular position encoders are:

1. The *tachometer encoder* or *rotary pulse generator* is a single-output device which generates pulses in response to sensor rotation. Only increments of motion are sensed, not direction. Fig. 10-10(a) depicts a rotary pulse generator which outputs a signal every time a magnet passes by. The angular resolution of this device in degrees is $360/n$, where n is the number of magnets.

If the device is energized or sampled for a fixed time interval, the number of output pulses per unit time represents angular velocity.

TABLE 10-2.   Rotary Position Sensors.

Device	Output	Accuracy
Multipole resolver	analog (voltage)	7 arc seconds
Absolute optical encoder	digital pulse	23 arc seconds
Standard resolver	analog (voltage)	7 arc minutes
Potentiometer	analog (resistance)	7 arc minutes
Incremental optical encoder	digital pulse	11 arc minutes
Contact encoder	digital pulse	26 arc minutes

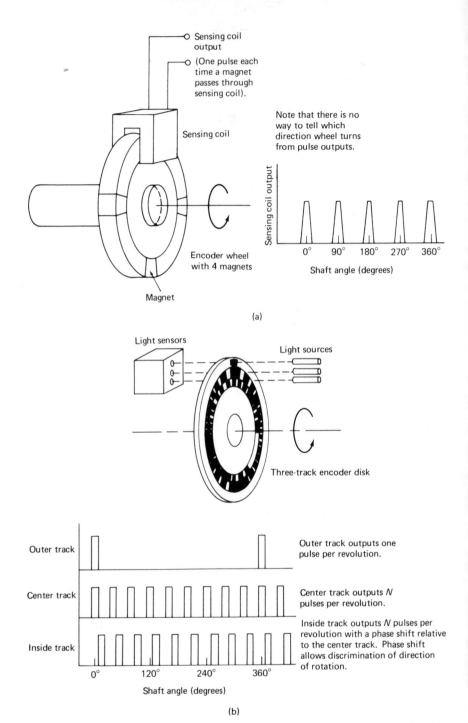

Fig. 10-10. (a) Magnetic tachometer encoder. **(b)** Incremental optical encoder. **(c)** Four-bit absolute optical encoder.

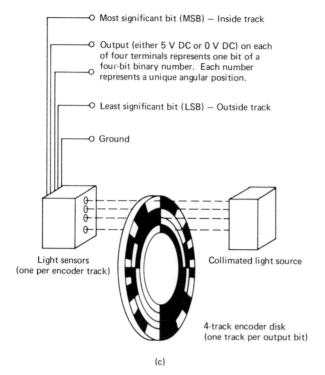

O Most significant bit (MSB) — Inside track

O Output (either 5 V DC or 0 V DC) on each
of four terminals represents one bit of a
four-bit binary number. Each number
O represents a unique angular position.

O Least significant bit (LSB) — Outside track

O Ground

Light sensors
(one per encoder track)

Collimated light source

4-track encoder disk
(one track per output bit)

(c)

Fig. 10-10. (*Continued*)

2. The *incremental encoder* is a three-output device which generates two of its outputs exactly as does the tachometer encoder. The two pulse trains, however, are out of phase with respect to each other so that *direction* of rotation as well as amount of angular rotation can be determined.

The third output is usually a single pulse once every full revolution. Fig. 10-10(b) depicts the widely used optical incremental encoder which generates pulses as alternating light and dark falls on a phototransistor.

In order to determine exact angular position at any given time, the output pulses must be counted, up for each increment in one direction, and down for each increment in the other. The total pulse count represents total angular displacement from the start up position.

3. *Absolute encoder* has 6–20 outputs, each of which is one bit of the binary representation of encoder shaft angular position. There is a one-for-one relationship between each specific angular position within one revolution and each binary number read at the outputs. Fig. 10-10(c) depicts an absolute encoder with four outputs, each of which requires a track on the encoder disk.

Three techniques of generating the encoder pulse outputs are generally used: direct contact, magnetic and optical.

Direct contact encoders use brushes which make electrical contact with a pattern of alternating conductive and nonconductive segments on the rotating encoder disk. Each encoder track switches output voltage ON and OFF as the brush contacts conducting and nonconducting segments, respectively. One advantage of brush-type encoders is that output voltage (i.e., the voltage level of pulses) can be simply adjusted by changing the input voltage level. In this manner, the encoder can easily be adapted to any type of interfacing logic circuitry. Disadvantages of brush-type encoders include brush friction, brush wear, and brush arcing.

Magnetic angular encoders use disks with ferromagnetic materials applied as a coating. In much the same way as data is encoded on magnetic tape by alternate polarity magnetic domains, the angular pattern is recorded on each encoder track. Small sensing coils are used to detect the passing of magnetized domains as the encoder disk rotates. These sensors are usually toroidal transformers excited with a high-frequency (200 kHZ) signal. A magnetized area passing by the coil results in a low (i.e., logical 0) signal on the coil output winding. Obviously these types of encoders require support electronics and are more complex than contact encoders. They are, however, rugged and reliable and can be used in harsh environments.

The most widely used encoders are of the optical type. The basic encoder element is the rotating disk, which is usually glass, and which has a precise, photographically imprinted pattern of alternating transparent and opaque segments. A light beam (LED or incandescent) passing through the disk at a single radius will energize photoelectric sensors at the same radius (but behind the disk) in a pattern uniquely related to disk (thus shaft) angular position. The photosensor outputs are usually too small to be used directly, so the pulses are electronically amplified and shaped to give square wave outputs compatible with the interfacing logic system.

The resolution of an absolute encoder of any type depends on the number of different angular segments or "slots" on the least-significant-bit track. In order to read out a binary number of the same resolution, there must be n tracks such that $2^n = S$, where S is the number of slots in the least-significant-bit track.

Absolute encoders are usually referred to as n-bit encoders and will have an output terminal for each bit. Table 10-3 shows the resolution in degrees associated with absolute encoders of various outputs.

Encoders are not control devices which can be used directly in machine sequencing systems in the same way limit switches are used; however, they can be used in conjunction with a *counter* to switch machine power or control loads. A counter is a device, usually solid state, which can be programmed to

TABLE 10-3. Absolute Encoder Resolution.

Absolute Encoder Bits	Resolution
6	5.6 degrees
8	1.4 degrees
10	0.4 degrees
12	5.3 minutes
14	1.3 minutes
16	0.3 minutes
18	4.9 seconds
20	1.2 seconds

switch machine power loads when a series of input pulses reaches a predetermined amount. For a machine subsystem which is driven by rotating elements, a *programmable limit switch* can then be made by combining an incremental encoder and a programmable timer. The encoder outputs a train of pulses as it rotates to represent linear motion of a machine element through some rotary-to-linear motion device (lead screw, rack and pinion, etc.). The counter counts the pulses, and switches at a preset number of pulses, which in fact represents a particular linear position of the machine element.

10.5.2. Analog Angular Position Sensors

Encoders described in Section 10.5.1 are inherently digital angular position devices. There are, however, several commercially available analog output sensors, including *potentiometers, synchros,* and *resolvers.*

Rotary potentiometers are devices which exhibit an output resistance which is linearly proportional to shaft angle inputs. Common ranges are 1 turn (360°), 3 turns (1080°), 5 turns (1800°), and 10 turns (3600°). Resistance is varied as a slidewire moves along a precision wire-wound resistor over the range of potentiometer travel.

Potentiometers have resolutions on the order of 1° and mechanical lifetimes in the range of 2–10 million revolutions. Models are available with ball bearing shaft support, capable of turning at relatively high speed, for use in servo positioning systems.

A synchro is a rotating transformer wherein a single-pole rotor winding is excited by AC voltage (60 Hz or 400 Hz). AC rotor current induces voltage in three stator coil windings which are wound 120° apart. The synchro is not different in appearance from a small AC motor.

Figure 10-11 shows the basic construction on a synchro and its circuit representation. If a voltage excitation of $A \sin \omega t$ is applied to the synchro

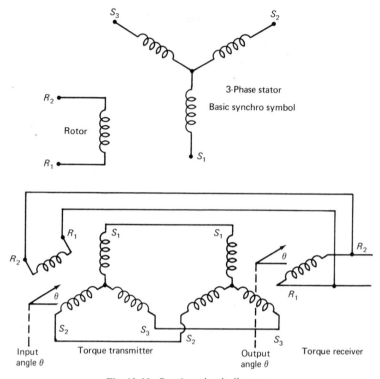

Fig. 10-11. Synchro circuit diagram.

rotor, then the output which will appear at the various stator terminals will be:

$$S_{13} = A \sin \omega t \sin \theta$$

$$S_{32} = A \sin \omega t \sin (\theta + 120°)$$

$$S_{21} = A \sin \omega t \sin (\theta + 240°)$$

where θ is the shaft angle of the synchro.

One of the unique features of a synchro is that two identical devices may be connected remotely, stator terminals to stator terminals and rotor terminals to rotor terminals to excitation voltage source, to create a "torque chain." As the input shaft is turned on the first synchro (called the *torque transmitter*), the shaft on the other synchro (called the *torque receiver*) turns through an equal angle.

Torque synchros are accurate to $\pm 1°$ and were once widely used to transmit shaft position to remote indicators in aircraft or ship navigation instruments.

The amount of torque which can be transmitted is limited by the amount of current which can flow through the synchro windings, and is relatively small.

Control synchros are used in a "control chain" as are torque synchros, but instead of exciting the output synchro rotor, it is held fixed and the output voltage measured. The output voltage varies from maximum to zero to maximum as the shaft of the input synchro (control transmitter) is rotated through 360°.

Resolvers are a form of synchro, in which there are only two stator windings, placed 90° mechanically out of phase with each other. If the rotor is excited with a voltage $A \sin \omega t$, then the outputs on stator terminals will be

$$S_{13} = A \sin \omega t \sin \theta$$

$$S_{12} = A \sin \omega t \cos \theta$$

where θ is the resolver shaft angle.

In order for synchros or resolvers to be used in machine control systems, some support electronics are necessary. Zero crossing detection and pulse shaping is generally used to result in an output signal similar to encoder outputs.

APPENDIX I

The following motor diagrams and dimensional specifications are reprinted from the "ANSI/NEMA STANDARDS PUBLICATION/NO. MG1-1978" (Revision 6—June 1981) published by the National Electrical Manufacturers Association, Washington, D.C. 20037.

NEMA Figures 4-1 and 4-2 define the dimensions of foot-mounted AC and DC motors. Tables MG 1-11.31 and MG 1-11.61 identify the actual standard dimensions for AC and DC motors respectively, as a function of motor frame designation. Thus the machine designer may design a machine to accommodate a motor of known frame size, without having selected the specific motor to be used.

NEMA Figure 4-3 defines the dimensions of face-mounted motors and Table MG 1-11.35 identifies the dimensions as a function of frame designation.

NEMA Figure 4-4 defines the dimensions of flange mounted motors and Table MG 1-11.37 identifies the dimensions as a function of frame designation.

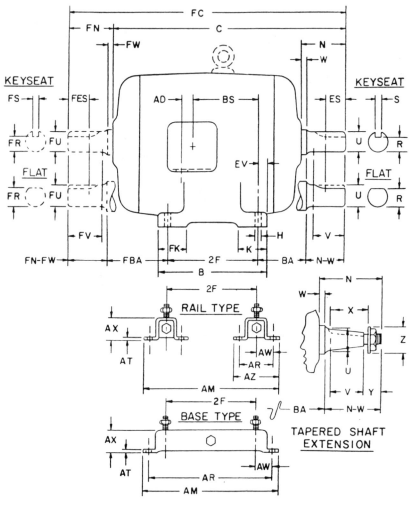

Fig. 4-1. Lettering of Dimension Sheets for Foot-mounted Machines—Side View.

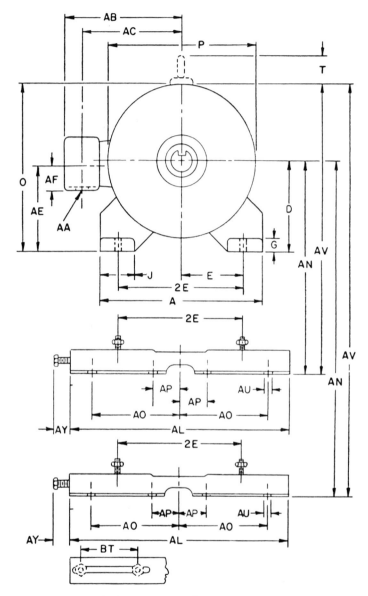

Fig. 4-2. Lettering of Dimension Sheets for Foot-mounted Machines—Drive End View.

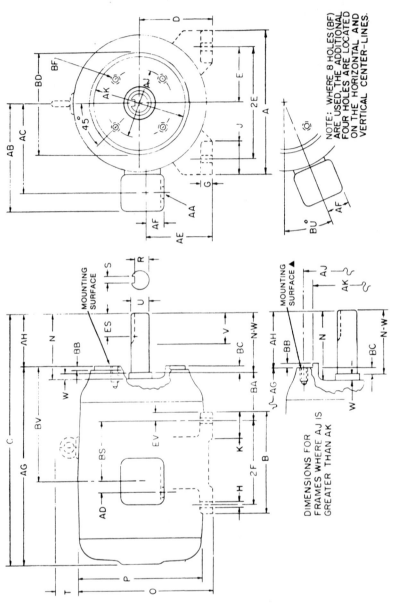

Fig. 4-3. Lettering of Dimension Sheets for Type C Face-mounted Foot or Footless Machines.

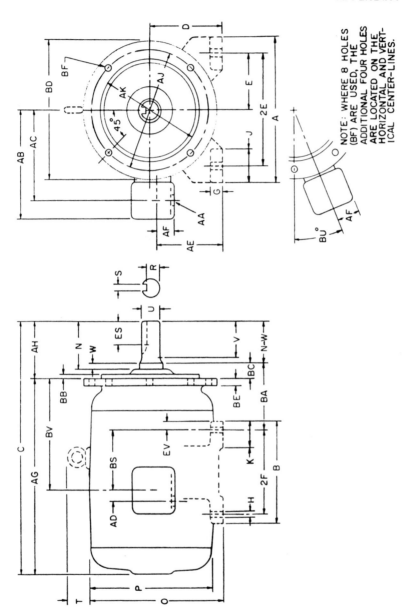

NOTE: WHERE 8 HOLES (BF) ARE USED, THE ADDITIONAL FOUR HOLES ARE LOCATED ON THE HORIZONTAL AND VERTICAL CENTER-LINES.

Fig. 4-4. Lettering of Dimension Sheets for Type D Flange-mounted Foot or Footless Machines.

MG 1-11.31 Dimensions for Alternating-current Foot-mounted Motors and Generators with Single Straight-shaft Extension

Frame Designation	A Max	B Max	D*	E†	2F‡	BA	H‡	U	N-W	V Min	R	Keyseat ES Min	Keyseat S	AA Min‡
42	…	…	2.62	1.75	1.69	2.06	0.28 slot	0.3750	1.12	…	0.328	…	flat	…
48	…	…	3.00	2.12	2.75	2.50	0.34 slot	0.5000	1.50	…	0.453	…	flat	…
48H	…	…	3.00	2.12	4.75	2.50	0.34 slot	0.5000	1.50	…	0.453	…	flat	…
56	…	…	3.50	2.44	3.00	2.75	0.34 slot	0.6250	1.88	…	0.517	1.41	0.188	…
56H	…	…	3.50	2.44	5.00	2.75	0.34 slot	0.6250	1.88	…	0.517	1.41	0.188	…
143T	7.0	6.0	3.50	2.75	4.00	2.25	0.34 hole	0.8750	2.25	2.00	0.771	1.41	0.188	3/4
145T	7.0	6.0	3.50	2.75	5.00	2.25	0.34 hole	0.8750	2.25	2.00	0.771	1.41	0.188	3/4
182T	9.0	6.5	4.50	3.75	4.50	2.75	0.41 hole	1.1250	2.75	2.50	0.986	1.78	0.250	3/4
184T	9.0	7.5	4.50	3.75	5.50	2.75	0.41 hole	1.1250	2.75	2.50	0.986	1.78	0.250	3/4
213T	10.5	7.5	5.25	4.25	5.50	3.50	0.41 hole	1.3750	3.38	3.12	1.201	2.41	0.312	1
215T	10.5	9.0	5.25	4.25	7.00	3.50	0.41 hole	1.3750	3.38	3.12	1.201	2.41	0.312	1
254T	12.5	10.8	6.25	5.00	8.25	4.25	0.53 hole	1.625	4.00	3.75	1.416	2.91	0.375	1 1/4
256T	12.5	12.5	6.25	5.00	10.00	4.25	0.53 hole	1.625	4.00	3.75	1.416	2.91	0.375	1 1/4
284T	14.0	12.5	7.00	5.50	9.50	4.75	0.53 hole	1.875	4.62	4.38	1.591	3.28	0.500	1 1/2
284TS	14.0	12.5	7.00	5.50	9.50	4.75	0.53 hole	1.625	3.25	3.00	1.416	1.91	0.375	1 1/2
286T	14.0	14.0	7.00	5.50	11.00	4.75	0.53 hole	1.875	4.62	4.38	1.591	3.28	0.500	1 1/2
286TS	14.0	14.0	7.00	5.50	11.00	4.75	0.53 hole	1.625	3.25	3.00	1.416	1.91	0.375	1 1/2
324T	16.0	14.0	8.00	6.25	10.50	5.25	0.66 hole	2.125	5.25	5.00	1.845	3.91	0.500	2
324TS	16.0	14.0	8.00	6.25	10.50	5.25	0.66 hole	1.875	3.75	3.50	1.591	2.03	0.500	2
326T	16.0	15.5	8.00	6.25	12.00	5.25	0.66 hole	2.125	5.25	5.00	1.845	3.91	0.500	2
326TS	16.0	15.5	8.00	6.25	12.00	5.25	0.66 hole	1.875	3.75	3.50	1.591	2.03	0.500	2
364T	18.0	15.2	9.00	7.00	11.25	5.88	0.66 hole	2.375	5.88	5.62	2.021	4.28	0.625	3
364TS	18.0	15.2	9.00	7.00	11.25	5.88	0.66 hole	1.875	3.75	3.50	1.591	2.03	0.500	3
365T	18.0	16.2	9.00	7.00	12.25	5.88	0.66 hole	2.375	5.88	5.62	2.021	4.28	0.625	3
365TS	18.0	16.2	9.00	7.00	12.25	5.88	0.66 hole	1.875	3.75	3.50	1.591	2.03	0.500	3
404T	20.0	16.2	10.00	8.00	12.25	6.62	0.81 hole	2.875	7.25	7.00	2.450	5.65	0.750	3
404TS	20.0	16.2	10.00	8.00	12.25	6.62	0.81 hole	2.125	4.25	4.00	1.845	2.78	0.500	3
405T	20.0	17.8	10.00	8.00	13.75	6.62	0.81 hole	2.875	7.25	7.00	2.450	5.65	0.750	3
405TS	20.0	17.8	10.00	8.00	13.75	6.62	0.81 hole	2.125	4.25	4.00	1.845	2.78	0.500	3
444T	22.0	18.5	11.00	9.00	14.50	7.50	0.81 hole	3.375	8.50	8.25	2.880	6.91	0.875	3
444TS	22.0	18.5	11.00	9.00	14.50	7.50	0.81 hole	2.375	4.75	4.50	2.021	3.03	0.625	3
445T	22.0	20.5	11.00	9.00	16.50	7.50	0.81 hole	3.375	8.50	8.25	2.880	6.91	0.875	3
445TS	22.0	20.5	11.00	9.00	16.50	7.50	0.81 hole	2.375	4.75	4.50	2.021	3.03	0.625	3

All dimensions in inches.

* Dimension D will never be greater than the above values for rigid-base motors. However, it may be less, so that shims are usually required for coupled or geared machines. When the exact dimension is required, shims up to 0.03 inch may be necessary on frame sizes whose D dimension is 8.00 inches or less; on larger frames, shims up to 0.06 inch may be necessary. No tolerances have been established for the D dimension of resilient mounted motors.

† Frames 42, 48, 48H, 56 and 56H—The tolerance for the 2F dimension shall be ±0.03 inch and for the H dimension (width of slot) shall be +0.02 inch, −0 inch. Frames 143T to 445T, inclusive—The tolerance for the 2E and 2F dimensions shall be ±0.03 inch and for the H dimension shall be +0.05 inch, −0 inch.

‡ For dimensions of clearance holes, see MG 1-4.04.

NOTE I.—For the meaning of the letter dimensions, see MG 1-4.01 and Figs

NOTE II.—For tolerances on shaft extension diameters and keyseats, see MG 1-4.05.

NOTE III.—It is recommended that all machines with keyseats cut in the shaft extension for pulley, coupling, pinion, etc. be furnished with a key unless otherwise specified by the purchaser.

NOTE IV.—Frames 42, 48, 48H, 56, 56H—If the shaft extension length of the motor is not suitable for the application, a length-ended... that deviations from this length in 0.25-inch

MG 1-11.35   Dimensions for Type C Face-mounted Foot or Footless Alternating-current Motors

Frame Designation	AJ**	AK	BA	BB Min	BC	BD Max	Number	Tap Size	Bolt Penetration Allowance	U	AH	R	ES Min	S
42C	3.750	3.000	2.062	0.16†	−0.19	5.00‡	4	1/4–20	...	0.3750	1.312*	0.328	...	flat
48C	3.750	3.000	2.50	0.16†	−0.19	5.625‡	4	1/4–20	...	0.500	1.69*	0.453	...	flat
56C	5.875	4.500	2.75	0.16†	−0.19	6.50‡	4	3/8–16	...	0.6250	2.06*	0.517	1.41	0.188
143TC and 145TC	5.875	4.500	2.75	0.16†	+0.12	6.50‡	4	3/8–16	0.56	0.8750	2.12	0.771	1.41	0.188
182TC and 184TC	7.250	8.500	3.50	0.25	+0.12	9.00‡	4	1/2–13	0.75	1.1250	2.62	0.986	1.78	0.250
182TCH and 184TCH	5.875	4.500	3.50	0.16†	+0.12	6.50‡	4	3/8–16	0.56	1.1250	2.62	0.986	1.78	0.250
213TC and 215TC	7.250	8.500	4.25	0.25	+0.25	9.00	4	1/2–13	0.75	1.3750	3.12	1.201	2.41	0.312
254TC and 256TC	7.250	8.500	4.75	0.25	+0.25	10.00	4	1/2–13	0.75	1.625	3.75	1.416	2.91	0.375
284TC and 286TC	9.000	10.500	4.75	0.25	+0.25	11.25	4	1/2–13	0.75	1.875	4.38	1.591	3.28	0.500
284TSC and 286TSC	9.000	10.500	4.75	0.25	+0.25	11.25	4	1/2–13	0.75	1.625	3.00	1.416	1.91	0.375
324TC and 326TC	11.000	12.500	5.25	0.25	+0.25	14.00	4	5/8–11	0.94	2.125	5.00	1.845	3.91	0.500
324TSC and 326TSC	11.000	12.500	5.25	0.25	+0.25	14.00	4	5/8–11	0.94	1.875	3.50	1.591	2.03	0.500
364TC and 365TC	11.000	12.500	5.88	0.25	+0.25	14.00	8	5/8–11	0.94	2.375	5.62	2.021	4.28	0.625
364TSC and 365TSC	11.000	12.500	5.88	0.25	+0.25	14.00	8	5/8–11	0.94	1.875	3.50	1.591	2.03	0.500
404TC and 405TC	11.000	12.500	6.62	0.25	+0.25	15.50	8	5/8–11	0.94	2.875	7.00	2.450	5.65	0.750
404TSC and 405TSC	11.000	12.500	6.62	0.25	+0.25	15.50	8	5/8–11	0.94	2.125	4.00	1.845	2.78	0.500
444 TC and 445TC	14.000	16.000	7.50	0.25	+0.25	18.00	8	5/8–11	0.94	3.375	8.25	2.880	6.91	0.875
444TSC and 445TSC	14.000	16.000	7.50	0.25	+0.25	18.00	8	5/8–11	0.94	2.375	4.50	2.021	3.03	0.625
500 frame series	14.500	16.500	...	0.25	+0.25	18.00	4	5/8–11	0.94	...	...	...	...	...

All dimensions in inches.

* If the shaft extension length of the motor is not suitable for the application, it is recommended that divisions from the length be in 0.25-inch increments.

** For frames 182TC and 184TC, and 213TC through 500TC, the centerline of the bolt holes shall be within 0.025 inch of true location. True location is defined as angular and diametrical location with reference to the centerline of the AK dimension.

† These BB dimensions are maximum dimensions.
‡ These BD dimensions are nominal dimensions.

NOTE I—For the meaning of the letter dimensions, see MG 1-4.01 and Fig. 4-3.

NOTE II—For frames 42C to 445 TSC, see MG 1-11.31 for dimensions A, B, D, E, 2F and H. For the 500 frame series, see MG 1-11.33 for dimensions D, E, 2F and BA.

NOTE III—For tolerances on shaft extension diameters and keyseats, see MG 1-4.05.

NOTE IV—For tolerance on AK dimensions, face runout and permissible eccentricity of mounting rabbet, see MG 1-4.07.

MG 1-11.37 Dimensions for Type D Flange-mounted Foot or Footless Alternating-current Motors

| Frame Designations | AJ | AK | BA | BB* | BC | BD Max | BE Nom | BF Hole | | Recommended Bolt Length | U | AH | R | Keyseat | |
								Number	Size					ES Min	S
143TD and 145TD	10.00	9.000	2.75	0.25	0.00	11.00	0.50	4	0.53	1.25	0.8750	2.25	0.771	1.41	0.188
182TD and 184TD	10.00	9.000	3.50	0.25	0.00	11.00	0.50	4	0.53	1.25	1.1250	2.75	0.986	1.78	0.250
213TD and 215TD	10.00	9.000	4.25	0.25	0.00	11.00	0.50	4	0.53	1.25	1.3750	3.38	1.201	2.41	0.312
254TD and 256TD	12.50	11.000	4.75	0.25	0.00	14.00	0.75	4	0.81	2.00	1.625	4.00	1.416	2.91	0.375
284TD and 286TD	12.50	11.000	4.75	0.25	0.00	14.00	0.75	4	0.81	2.00	1.875	4.62	1.591	3.28	0.500
284TSD and 286TSD	12.50	11.000	4.75	0.25	0.00	14.00	0.75	4	0.81	2.00	1.625	3.25	1.416	1.91	0.375
324TD and 326TD	16.00	14.000	5.25	0.25	0.00	18.00	0.75	4	0.81	2.00	2.125	5.25	1.845	3.91	0.500
324TSD and 326TSD	16.00	14.000	5.25	0.25	0.00	18.00	0.75	4	0.81	2.00	1.875	3.75	1.591	2.03	0.500
364TD and 365TD	16.00	14.000	5.88	0.25	0.00	18.00	0.75	4	0.81	2.00	2.375	5.88	2.021	4.28	0.625
364TSD and 365TSD	16.00	14.000	5.88	0.25	0.00	18.00	0.75	4	0.81	2.00	1.875	3.75	1.591	2.03	0.500
404TD and 405TD	20.00	18.000	6.62	0.25	0.00	22.00	1.00	8	0.81	2.25	2.875	7.25	2.450	5.65	0.750
404TSD and 405TSD	20.00	18.000	6.62	0.25	0.00	22.00	1.00	8	0.81	2.25	2.125	4.25	1.845	2.78	0.500
444TD and 445TD	20.00	18.000	7.50	0.25	0.00	22.00	1.00	8	0.81	2.25	3.375	8.50	2.880	6.91	0.875
444TSD and 445TSD	20.00	18.000	7.50	0.25	0.00	22.00	1.00	8	0.81	2.25	2.375	4.75	2.021	3.03	0.625
500 frame series	22.00	18.000	...	0.25	0.00	25.00	...	8	0.81	...	...	...	...	...	...

All dimensions in inches.
* Tolerance is +0.00 inch, −0.06 inch.
NOTE I—For the meaning of the letter dimensions, see MG 1-4.01 and Fig. 4-4.
NOTE II—For frames 143TD–445TSD, see MG 1-11.31 for dimensions A, B, D, E, 2F and H. For the 500 frame series, see MG 1-11.33 for dimensions D, E, 2F and BA.
NOTE III—For tolerances on shaft extension diameters and keyseats, see MG 1-4.05.
NOTE IV—For tolerance on AK dimension, face runout and permissible eccentricity of mounting rabbet, see MG 1-4.07.

MG 1-11.61 Dimensions for Foot-mounted Industrial Direct-current Motors and Generators

Frame Designations	A Max	B Max	D*	E†	2F†	BA	H† Hole	AL	AM	AO	AR	AU	AX	AY Max Bases	BT
182AT	9.00	6.50	4.50	3.75	4.50	2.75	0.41	12.75	9.50	4.50	4.25	0.50	1.50	0.50	3.00
183AT	9.00	7.00	4.50	3.75	5.00	2.75	0.41	12.75	10.00	4.50	4.50	0.50	1.50	0.50	3.00
184AT	9.00	7.50	4.50	3.75	5.50	2.75	0.41	12.75	10.50	4.50	4.75	0.50	1.50	0.50	3.00
185AT	9.00	8.25	4.50	3.75	6.25	2.75	0.41	12.75	11.25	4.50	5.12	0.50	1.50	0.50	3.00
186AT	9.00	9.00	4.50	3.75	7.00	2.75	0.41	12.75	12.00	4.50	5.50	0.50	1.50	0.50	3.00
187AT	9.00	10.00	4.50	3.75	8.00	2.75	0.41	12.75	13.00	4.50	6.00	0.50	1.50	0.50	3.00
188AT	9.00	11.00	4.50	3.75	9.00	2.75	0.41	12.75	14.00	4.50	6.50	0.50	1.50	0.50	3.00
189AT	9.00	12.00	4.50	3.75	10.00	2.75	0.41	12.75	15.00	4.50	7.00	0.50	1.50	0.50	3.00
1810AT	9.00	13.00	4.50	3.75	11.00	2.75	0.41	12.75	16.00	4.50	7.50	0.50	1.50	0.50	3.00
213AT	10.50	7.50	5.25	4.25	5.50	3.50	0.41	15.00	11.00	5.25	4.75	0.50	1.75	0.50	3.50
214AT	10.50	8.25	5.25	4.25	6.25	3.50	0.41	15.00	11.75	5.25	5.12	0.50	1.75	0.50	3.50
215AT	10.50	9.00	5.25	4.25	7.00	3.50	0.41	15.00	12.50	5.25	5.50	0.50	1.75	0.50	3.50
216AT	10.50	10.00	5.25	4.25	8.00	3.50	0.41	15.00	13.50	5.25	6.00	0.50	1.75	0.50	3.50
217AT	10.50	11.00	5.25	4.25	9.00	3.50	0.41	15.00	14.50	5.25	6.50	0.50	1.75	0.50	3.50
218AT	10.50	12.00	5.25	4.25	10.00	3.50	0.41	15.00	15.50	5.25	7.00	0.50	1.75	0.50	3.50
219AT	10.50	13.00	5.25	4.25	11.00	3.50	0.41	15.00	16.50	5.25	7.50	0.50	1.75	0.50	3.50
2110AT	10.50	14.50	5.25	4.25	12.50	3.50	0.41	15.00	18.00	5.25	8.25	0.50	1.75	0.50	3.50
253AT	12.50	9.50	6.25	5.00	7.00	4.25	0.53	17.75	13.88	6.25	6.00	0.62	2.00	0.62	4.00
254AT	12.50	10.75	6.25	5.00	8.25	4.25	0.53	17.75	15.12	6.25	6.62	0.62	2.00	0.62	4.00
255AT	12.50	11.50	6.25	5.00	9.00	4.25	0.53	17.75	15.88	6.25	7.00	0.62	2.00	0.62	4.00
256AT	12.50	12.50	6.25	5.00	10.00	4.25	0.53	17.75	16.88	6.25	7.50	0.62	2.00	0.62	4.00
257AT	12.50	13.50	6.25	5.00	11.00	4.25	0.53	17.75	17.88	6.25	8.00	0.62	2.00	0.62	4.00
258AT	12.50	15.00	6.25	5.00	12.50	4.25	0.53	17.75	19.38	6.25	8.75	0.62	2.00	0.62	4.00
259AT	12.50	16.50	6.25	5.00	14.00	4.25	0.53	17.75	20.88	6.25	9.50	0.62	2.00	0.62	4.00
283AT	14.00	11.00	7.00	5.50	8.00	4.75	0.53	19.75	15.38	7.00	6.75	0.62	2.00	0.62	4.50
284AT	14.00	12.50	7.00	5.50	9.50	4.75	0.53	19.75	16.88	7.00	7.50	0.62	2.00	0.62	4.50
285AT	14.00	13.00	7.00	5.50	10.00	4.75	0.53	19.75	17.38	7.00	7.75	0.62	2.00	0.62	4.50
286AT	14.00	14.00	7.00	5.50	11.00	4.75	0.53	19.75	18.38	7.00	8.25	0.62	2.00	0.62	4.50
287AT	14.00	15.50	7.00	5.50	12.50	4.75	0.53	19.75	19.88	7.00	9.00	0.62	2.00	0.62	4.50
288AT	14.00	17.00	7.00	5.50	14.00	4.75	0.53	19.75	21.38	7.00	9.75	0.62	2.00	0.62	4.50
289AT	14.00	19.00	7.00	5.50	16.00	4.75	0.53	19.75	23.38	7.00	10.75	0.62	2.00	0.62	4.50
323AT	16.00	12.50	8.00	6.25	9.00	5.25	0.66	22.75	17.75	8.00	7.75	0.75	2.50	0.75	5.25
324AT	16.00	14.00	8.00	6.25	10.50	5.25	0.66	22.75	19.25	8.00	8.50	0.75	2.50	0.75	5.25
325AT	16.00	14.50	8.00	6.25	11.00	5.25	0.66	22.75	19.75	8.00	8.75	0.75	2.50	0.75	5.25
326AT	16.00	15.50	8.00	6.25	12.00	5.25	0.66	22.75	20.75	8.00	9.25	0.75	2.50	0.75	5.25
327AT	16.00	17.50	8.00	6.25	14.00	5.25	0.66	22.75	22.75	8.00	10.25	0.75	2.50	0.75	5.25
328AT	16.00	19.50	8.00	6.25	16.00	5.25	0.66	22.75	24.75	8.00	11.25	0.75	2.50	0.75	5.25
329AT	16.00	21.50	8.00	6.25	18.00	5.25	0.66	22.75	26.75	8.00	12.25	0.75	2.50	0.75	5.25

(Continued)

MG 1-11.61 (Continued)

Frame Designations	A Max	B Max	D*	E†	2F†	BA	Ht Hole	AL	AM	AO	AR	AU	AX	AY Max Bases	BT
363AT	18.00	14.00	9.00	7.00	10.00	5.88	0.81	25.50	19.25	9.00	8.25	0.88	2.50	0.75	6.00
364AT	18.00	15.25	9.00	7.00	11.25	5.88	0.81	25.50	20.50	9.00	9.12	0.88	2.50	0.75	6.00
365AT	18.00	16.25	9.00	7.00	12.25	5.88	0.81	25.50	21.50	9.00	9.62	0.88	2.50	0.75	6.00
366AT	18.00	18.00	9.00	7.00	14.00	5.88	0.81	25.50	23.25	9.00	10.50	0.88	2.50	0.75	6.00
367AT	18.00	20.00	9.00	7.00	16.00	5.88	0.81	25.50	25.25	9.00	11.50	0.88	2.50	0.75	6.00
368AT	18.00	22.00	9.00	7.00	18.00	5.88	0.81	25.50	27.25	9.00	12.50	0.88	2.50	0.75	6.00
369AT	18.00	24.00	9.00	7.00	20.00	5.88	0.81	25.50	29.25	9.00	13.50	0.88	2.50	0.75	6.00
403AT	20.00	15.00	10.00	8.00	11.00	6.62	0.94	28.75	21.12	10.00	9.25	1.00	3.00	0.88	7.00
404AT	20.00	16.25	10.00	8.00	12.25	6.62	0.94	28.75	22.38	10.00	9.88	1.00	3.00	0.88	7.00
405AT	20.00	17.75	10.00	8.00	13.75	6.62	0.94	28.75	23.88	10.00	10.62	1.00	3.00	0.88	7.00
406AT	20.00	20.00	10.00	8.00	16.00	6.62	0.94	28.75	26.12	10.00	11.75	1.00	3.00	0.88	7.00
407AT	20.00	22.00	10.00	8.00	18.00	6.62	0.94	28.75	28.12	10.00	12.75	1.00	3.00	0.88	7.00
408AT	20.00	24.00	10.00	8.00	20.00	6.62	0.94	28.75	30.12	10.00	13.75	1.00	3.00	0.88	7.00
409AT	20.00	26.00	10.00	8.00	22.00	6.62	0.94	28.75	32.12	10.00	14.75	1.00	3.00	0.88	7.00
443AT	22.00	16.50	11.00	9.00	12.50	7.50	1.06	31.25	22.62	11.00	10.00	1.12	3.00	0.88	7.50
444AT	22.00	18.50	11.00	9.00	15.00	7.50	1.06	31.25	24.62	11.00	11.00	1.12	3.00	0.88	7.50
445AT	22.00	20.50	11.00	9.00	16.50	7.50	1.06	31.25	26.62	11.00	12.00	1.12	3.00	0.88	7.50
446AT	22.00	22.00	11.00	9.00	18.00	7.50	1.06	31.25	28.12	11.00	12.75	1.12	3.00	0.88	7.50
447AT	22.00	24.00	11.00	9.00	20.00	7.50	1.06	31.25	30.12	11.00	13.75	1.12	3.00	0.88	7.50
448AT	22.00	26.00	11.00	9.00	22.00	7.50	1.06	31.25	32.12	11.00	14.75	1.12	3.00	0.88	7.50
449AT	22.00	29.00	11.00	9.00	25.00	7.50	1.06	31.25	35.12	11.00	16.25	1.12	3.00	0.88	7.50
502AT	25.00	17.50	12.50	10.00	12.50	8.50	1.19	35.00	24.50	12.50	10.75	1.25	3.50	⋯	8.00
503AT	25.00	19.00	12.50	10.00	14.00	8.50	1.19	35.00	26.00	12.50	11.50	1.25	3.50	⋯	8.00
504AT	25.00	21.00	12.50	10.00	16.00	8.60	1.19	35.00	28.00	12.50	12.50	1.25	3.50	⋯	8.00
505AT	25.00	23.00	12.50	10.00	18.00	8.50	1.19	35.00	30.00	12.50	13.50	1.25	3.50	⋯	8.00
506AT	25.00	25.00	12.50	10.00	20.00	8.50	1.19	35.00	32.00	12.50	14.50	1.25	3.50	⋯	8.00
507AT	25.00	27.00	12.50	10.00	22.00	8.50	1.19	35.00	34.00	12.50	15.50	1.25	3.50	⋯	8.00
508AT	25.00	30.00	12.50	10.00	25.00	8.50	1.19	35.00	37.00	12.50	17.00	1.25	3.50	⋯	8.00
509AT	25.00	33.00	12.50	10.00	28.00	8.50	1.19	35.00	40.00	12.50	18.50	1.25	3.50	⋯	8.00
583A	29.00	21.00	14.50	11.50	16.00	10.00	1.19	38.75	29.00	14.50	13.00	1.25	4.00	⋯	8.50
584A	29.00	23.00	14.50	11.50	18.00	10.00	1.19	38.75	31.00	14.50	14.00	1.25	4.00	⋯	8.50
585A	29.00	25.00	14.50	11.50	20.00	10.00	1.19	38.75	33.00	14.50	15.00	1.25	4.00	⋯	8.50
586A	29.00	27.00	14.50	11.50	22.00	10.00	1.19	38.75	35.00	14.50	16.00	1.25	4.00	⋯	8.50
587A	29.00	30.00	14.50	11.50	25.00	10.00	1.19	38.75	38.00	14.50	17.50	1.25	4.00	⋯	8.50
588A	29.00	33.00	14.50	11.50	28.00	10.00	1.19	38.75	41.00	14.50	19.00	1.25	4.00	⋯	8.50
683A	34.00	25.00	17.00	13.50	20.00	11.50	1.19	42.50	30.75	13.50	14.00	1.38	4.25	⋯	9.00
684A	34.00	27.00	17.00	13.50	22.00	11.50	1.19	42.50	32.75	13.50	15.00	1.38	4.25	⋯	9.00
685A	34.00	30.00	17.00	13.50	25.00	11.50	1.19	42.50	35.75	13.50	16.50	1.38	4.25	⋯	9.00
686A	34.00	33.00	17.00	13.50	28.00	11.50	1.19	42.50	38.75	13.50	18.00	1.38	4.25	⋯	9.00
687A	34.00	37.00	17.00	13.50	32.00	11.50	1.19	42.50	42.75	13.50	20.00	1.38	4.25	⋯	9.00
688A	34.00	41.00	17.00	13.50	36.00	11.50	1.19	42.50	46.75	13.50	22.00	1.38	4.25	⋯	9.00

(Continued)

MG 1-11.61 (Continued)

Frame Designations‡	Drive End—For Belt Drive						Drive End—For Direct-connected Drive‡						End Opposite Drive—Straight					
					Keyseat						Keyseat						Keyseat	
	U	N–W	V Min	R	ES Min	S	U	N–W	V Min	R	ES Min	S	FU	FN–FW	FV Min	FR	FES Min	FS
182AT-1810AT	1.1250	2.25	2.00	0.986	1.41	0.250	…	…	…	…	…	…	0.8750	1.75	1.50	0.771	0.91	0.188
213AT-2110AT	1.3750	2.75	2.50	1.201	1.78	0.312	…	…	…	…	…	…	1.1250	2.25	2.00	0.986	1.41	0.250
253AT-259AT	1.625	3.25	3.00	1.416	2.28	0.375	…	…	…	…	…	…	1.3750	2.75	2.50	1.201	1.78	0.312
283AT-289AT	1.875	3.75	3.50	1.591	2.53	0.500	…	…	…	…	…	…	1.625	3.25	3.00	1.416	2.28	0.375
323AT-329AT	2.125	4.25	4.00	1.845	3.03	0.500	…	…	…	…	…	…	1.875	3.75	3.50	1.591	2.53	0.500
363AT-369AT	2.375	4.75	4.50	2.021	3.53	0.625	…	…	…	…	…	…	2.125	4.25	4.00	1.845	3.03	0.500
403AT-409AT	2.625	5.25	5.00	2.275	4.03	0.625	…	…	…	…	…	…	2.375	4.75	4.50	2.021	3.53	0.625
443AT-449AT	2.875	5.75	5.50	2.450	4.53	0.750	…	…	…	…	…	…	2.625	5.25	5.00	2.275	4.03	0.625
502AT-509AT	3.250	6.50	6.25	2.831	5.28	0.750	…	…	…	…	…	…	2.875	5.75	5.50	2.450	4.53	0.750
583A-588A	3.250	9.75	9.50	2.831	8.28	0.750	2.875	5.75	5.50	2.450	4.28	0.750	…	…	…	…	…	…
683A-688A	3.625	10.88	10.62	3.134	9.53	0.875	3.250	6.50	6.25	2.831	5.03	0.750	…	…	…	…	…	…

All dimensions in inches.

* Dimension D will never be greater than the values shown in the tables, but it may be less so that shims are usually required for coupled or geared machines. When the exact dimension is required, shims up to 0.03 inch may be necessary on frame sizes whose dimension D is 8 inches or less; on larger frames, shims up to 0.06 inch may be necessary.
† The tolerance for the 2E and 2F dimensions shall be ±0.03 inch and for the H dimension shall be ±0.05 inch, −0.0 inch.
‡ When frames 583A through 688A have a shaft extension for direct-connected drive, the frame number shall have the suffix letter "S" (e.g., 583AS).

NOTE I—For the meaning of the letter dimensions, see MG 1-4.01 and Figs. 4-1 and 4-2.
NOTE II—For tolerances on shaft extension diameters and keyseats, see MG 1-4.05.
NOTE III—It is recommended that all machines with keyseats cut in the shaft extension for pulley, coupling, pinion, etc., be furnished with a key unless otherwise specified by the purchaser.

INDEX

actuators 64, 70 (table)
   cylinders 128
   rotary 144, 145 (fig), 193
   solenoid 147, 148 (fig)
adhesive
   application force 96
   application machine 183
approach 2
automatic test equipment 12, 13
automation 5

Boolean algebra 265–272
   AND 269 (fig)
   identities 272 (table)
   NAND 270 (fig)
   NOR 271 (fig)
   NOT 268 (fig)
   OR 269 (fig)
   YES 267 (fig)
brainstorming 30

control components (see also sensors)
   electromechanical timers 303
     time delay 304
     reset timer 304
     repeat cycle timer 304
   limit switches 247, 296
     impulse limit switch 259, 300
   program timer 232 (fig)
   relays 239, 240 (fig)
     characteristics 243 (table)
     reed relay 242, 301
     solid state relay 242, 283, 284 (fig)
     time delay relay 242, 304
   rotary position sensors 311 (table)
     encoders 311
     potentiometers 315
     resolvers 315
     synchros 315

solenoid valves 138, 247
stepping switch 233
control system 9, 47, 69, 230–289
   algorithm 69
   control logic 69, 233, 265–284
   control scheme 236–238, 237 (fig)
   electronic logic 282
     CMOS 282
     example 284
     interfacing 283
     TTL 282
   generalized system 234, 235 (fig)
   monitoring function 234
   power requirements 44, 238
   protection function 233
   relay controls 239, 242–264
   sequencing 231, 254
     diagram 259
     event based 233
     load 254
     position 254
   synchronization 27
   system requirements 230, 233, 236 (table)
conveyor 74
creativity 29, 45
cylinders (see also fluid power)
   hydraulic 135
   pneumatic 129
   reciprocation example 248, 250, 251, 252
   sequencing example 255, 256, 258, 261

design
   alternatives 34, 43
   constraints 45 (table), 48
   method 31
     automatic machinery 34
     modular design 154

design (*Continued*)
  processes 3, 29–52
    system design 32
    table 30
  requirements 31, 38, 40, 44, 48 (table)
  synthesis 35, 45
design and build (D&B)
  business 19
  design review 25
  project 21
  proposal 23, 36
detail design 25, 34, 35, 51 (table)
  drawings 25
dispensing force 95
drive systems 64, 67
  design 71 (fig)
  transmission 68

encoder
  absolute 312 (fig), 313
  direct contact 314
  incremental 123, 312 (fig), 313
  magnetic 314
  optical 314
  rotary pulse generator 311, 312 (fig)
escapements 219–225
  control 224
  displacement 223
  gate 223
  jaw 224 (fig)
  ratchet 220, 221 (fig)
  rotary 222, 223 (fig)
  slide 221, 222 (fig)

fastening 159
feasibility study 24
flow 66
  in cylinders 136
fluid motors 142–144
  air 142
  hydraulic 144
fluid power 128–146
  actuators 128
  air jet 145
  cylinders 128, 130 (fig)
    configuration 132, 137
    force requirements 129, 136
    hydraulic 135
    mounting 134, 135 (fig), 137
    pneumatic 129
    rod size 134, 138

  speed requirements 131, 133 (fig), 136
  stroke requirements 131, 136
directional control valves 138, 139 (fig)
  actuators 140, 143
  diagrams 141 (fig)
  ports 140
  positions 140
  symbols 143 (fig)
  "ways" 140
  rotary actuators 144, 145 (fig)
  rotary tables 144
  shock absorber 145
  vacuum source 145 (fig)
friction
  bearing 75, 76 (table)
  coefficient 74 (table)
  coulomb 72
  rotary coulomb 75
  torque 75, 97
  viscous 77
functional visualization 30

gear reducer 83
generalized machine 6–9, 7 (fig)

heat build-up 76

ideation 30
inertial forces, torques 79
inspiration 31

Karnaugh mapping 272–282
  control application example 276
  four variable 275 (fig)
  three variable 274 (fig), 275 (fig)
  two variable 273 (fig)

limit switches (see switches)

machine
  build 26
  components 33
    off-the-shelf 8, 155 (table)
  concepting 22, 34, 35 (fig), 50
    alternate mechanizations 43
    concept evaluation 47, 49 (table)
    concept selection 49
    concept synthesis 45
    information requirements 36, 37 (fig)
    process requirements 38, 39 (table), 42
      (fig)

D&B contracts
firm fixed price   54
cost reimbursement (CPFF)   54
time and materials   54
costing   23, 25, 56
debug   26, 27
loads   64, 69, 70, 96
output   68
proof   26, 28
requirements   64, 96, 167, 169
fail safe   169
maintainability   169, 171 (table)
reliability   169, 171 (table)
safety   167, 170
setup   26
machine types   9–18
adaptive   5
assembly   9, 10, 11 (fig)
automatic   5
inspection   9, 11, 12 (fig)
machine tool   10, 15 (fig)
manual   5
packaging   10, 13, 14 (fig)
semi-automatic   5
test   10, 12, 14 (fig)
machinery design   1
machinery designer   2
machinery economics   53–63
costing   56
machining operations   89 (table)
drilling forces   91
speeds   89, 90
spindle efficiency   90
tool feed   90
unit power   89, 92 (table)
magnetic field
AC rotor   106 (fig), 107
DC motor field   110, 111 (fig)
rotating   103–105, 104 (fig)
manufacturing processes
batch   231
continuous   230
discrete component   231
material transfer   6, 8, 16, 173–229 (see also
work transfer; parts feeding)
carousel   16, 176 (fig), 181, 182 (fig)
continuous   16, 17. 174, 175 (fig), 179
cross-line   6, 9, 199–229
dial   16
flexible   18, 174 (fig), 176, 179
free transfer   18

inline   8, 174 (fig), 177, 186
intermittant   16, 17, 175
lift and carry   177
rotary   180, 181 (fig)
mainline   6, 8, 173–199
requirements   40
rotary   8, 16, 175 (fig), 180, 181, 192
walking beam   177, 190 (fig)
x-y table   183, 196 (fig)
metal forming   85
bending   85, 86 (fig)
channel bend   85, 87
edge bend   85, 87
V bend   85, 87
metal removal   87, 89
metal shearing   87
forces   87
microprocessor, microcomputer
control   287
modular design   154
moment of inertia
reflected   80, 83, 86 (fig), 97
table   81
morphological charts   30
motors, electric   97–128
accelerating torque   101, 102
breakaway torque   101
control   97, 100
dimensions APP I
duty cycle   97
enclosures   98 (table)
failure   98
frame sizes   98, APP I
inertia, rotor   102
loads   101, 102
NEMA standards   97
poles   103
ratings   97
rated load   99, 102
running torque   102
selection   97, 100
sizing   101, APP I
starting torque   103
stator   103, 103 (fig)
torque-speed curve   101
types   99, 100 (fig)
motors, AC   99, 103
capacitor start   105
induction   107
characteristics .109
breakdown torque   109

motors, AC (*Continued*)
  induction (*Continued*)
    NEMA designs 110
    pull up torque 109
    rated load 110
    slip 108
    squirrel cage rotor 107 (fig)
    starting torque 109
    torque-speed curve 109 (fig)
  motor speed 106
  rotating magnetic field 103
  single phase 105
  synchronous 107
  synchronous speed 106
  wound rotor 107
motors, DC 82, 99, 110, 111 (fig)
  armature 111
  armature current 112
  back EMF 112
  commutator 111
  compound 115
    characteristics 117 (fig)
    cumulative 118
    differential 118
  series 115
  shunt 115
  speed control 116
  torque-speed curve 115, 116 (fig)
motors, servo 124-128
  AC 124
  controls 125 (fig)
  DC 124
  permanent magnet field 125
  torque requirements 127
  velocity 126
motors, step 118-123
  applications 122
  bifilar 119
  controls 120, 121 (fig)
  half stepping 120
  indexer 122
  operation 118, 119 (fig)
  torque-speed curve 123, 124 (fig)
  translator 121

non-technical design factors 49

objectives 2
operations sheet 36, 40
organization 2

parts feeding 9, 200
parts feeders 200-213
  drum selector 208 (fig)
  elevatot 210, 212 (fig)
  external gate hopper 209 (fig)
  oscillating box 210, 211 (fig)
  rotary disk 207 (fig)
  rotating bowl 206-209
  stationary bowl 213-214
    centerboard feeders 214 (fig)
    reciprocating tube feeder 209, 211 (fig)
  tumbling barrel 207 (fig)
  vibratory bowl 201 (fig), 201-206
parts handling 199-229
  bulk handling 199
  off line manufacture 200
  on line manufacture 199
parts orientors 213-215
  design rules 215
  in-bowl tooling 213, 214 (fig)
parts placement 225-229
  pick and place 225, 226 (fig), 227 (fig)
parts presence 11
parts transfer 215
  design rules 216
  feed tracks 215, 216 (fig)
photoelectric switches (see switches)
plant visit 36, 38
pick and place 189, 191 (fig), 225
  grippers 229
  linear parts handler 226 (fig)
  rotary parts handler 227 (fig)
power 64
  equations 67 (table)
  control 44, 238
  fluid 66, 67 (fig), 128-146
precedence diagram 41, 42 (fig)
preliminary design 24
  layout 34, 35, 50
pressure 66
  cylinder 129, 131
prime mover 64
production 4
  manufacturing processes 230
programmable controllers 285, 286 (fig)
project costing 56, 61 (fig)
  direct costs 60
  direct labor 57, 58 (fig)
  direct materials 60
  general & administrative (G&A) 61

G&A pool 61
  indirect expenses 61
manpower loading 58 (fig)
overhead 59
overhead pool 59
overhead rate 59
profit 61
proximity switch (see switches)

relay logic 242–264 (see also control system)
  cylinder reciprocation examples 248, 250, 251, 252
  cylinder sequencing examples 225, 256, 258, 260, 261
  holding circuit 244, 248 (fig)
  ladder diagrams 243, 249, 250, 251, 253, 256, 257, 258, 259, 264
  symbols 245 (fig)
request for proposal RFP 24, 36, 39
request for quote RFQ 21
robots 4
  pick and place 225
rotary table 144, 180, 192, 193 (fig), 194 (fig), 195 (fig)

safety requirements 167
  considerations 169
  design for 170 (table)
  fail safe 169
sensors 290–317 (see also control components)
  continuous 290
  event 290
  linear position 290–311
    limit switches 296
    photoelectric switches 308
    proximity switches 306
  rotary position 311–317
shear strength 88 (table)
solenoid 147–153, 148 (fig)
  armature 147
  control circuits 151, 153 (fig)
  duty cycle 150
  holding force 147, 149 (fig)
  selection 150
  speed 153
  stroke 147
  temperature rise 151, 153 (fig)
special machines 4, 32

switches 291
  contacts 293, 294 (fig)
  electromechanical 291
  inductive loads 292
  limit switches 296
    actuators 298 (fig), 299 (fig), 300 (fig)
    enclosures 297 (table)
    impulse 259, 300
    manual 296
  photoelectric 308
    direct scan 308, 309 (fig)
    excess gain 310
    reflective 309 (fig)
    retro-reflective 309 (fig), 310
  pressure 302
    example 303 (fig)
  proximity 306
    capacitive 307
    Hall effect 308
    inductive 306
    magnetic 307
    ring type 308
    ultrasonic 308
  reed 301 (fig)
  solid state 305
synectics 30
syringe 95
system design 32–34, 33 (fig)
  optimal 32

tensile strength 88 (table)
tolerancing 25
tools 69
  dull tool 87, 95
  feed rate 90
torque 64, 97
  accelerating 80, 102
  electric motor 101
  induction motor 108
  inertial 79

unit power 89

vacuum
  chuck 164
  pickup 229
  source 146 (fig)
value engineering 30
vibratory bowl feeder 201–206, 201 (fig)
  bowl acceleration 205
  bowl load 206

vibratory bowl feeder (*Continued*)
    forces on parts 202–204
    frequency 205
    operation 201
    track angle 205
    track friction 206
    vibration angle 205

work carrier 69, 185
work holding 47, 177, 185–186
    nests 180
    work carrier 185
work interface 68
workpiece 64
    clamping 158, 164
    inertia 64
    positioning 157, 163
work station 6, 156, 163
    accuracy 165
    adjustment 166
    fastening 159
    off-the-shelf 7, 8, 155 (table)

    "packaged" 7, 154
    requirements 40, 156
        assembly machines 157
        inspection station 160
        machining station 161
        packaging station 162
        test station 161
work transfer 173–229 (see also material transfer)
    belt, friction drive 198 (fig)
    crossover cam 95 (fig)
    drive principles 186
        inline 186
        lift and carry 188, 190 (fig)
        pawl actuated 187 (fig)
        rotary drives 192
    geneva mechanism 194

x-y table 183, 196 (fig)
    design 196
    error 197